Wireless Networks

Series Editor

Xuemin Sherman Shen, University of Waterloo, Waterloo, Canada

The purpose of Springer's Wireless Networks book series is to establish the state of the art and set the course for future research and development in wireless communication networks. The scope of this series includes not only all aspects of wireless networks (including cellular networks, WiFi, sensor networks, and vehicular networks), but related areas such as cloud computing and big data. The series serves as a central source of references for wireless networks research and development. It aims to publish thorough and cohesive overviews on specific topics in wireless networks, as well as works that are larger in scope than survey articles and that contain more detailed background information. The series also provides coverage of advanced and timely topics worthy of monographs, contributed volumes, textbooks and handbooks.

Jie Xu · Xiaohua Jia

Machine Unlearning

Theory and Applications in Networking

Springer

Jie Xu
Department of Computer Science
City University of Hong Kong
Kowloon, Hong Kong

Xiaohua Jia
Department of Computer Science
City University of Hong Kong
Kowloon, Hong Kong

ISSN 2366-1186 ISSN 2366-1445 (electronic)
Wireless Networks
ISBN 978-3-032-30977-8 ISBN 978-3-032-30978-5 (eBook)
https://doi.org/10.1007/978-3-032-30978-5

This Springer imprint is published by the registered company Springer Nature Switzerland AG
The registered company address is: Gewerbestrasse 11, 6330 Cham, Switzerland

This book marks my first steps into artificial intelligence, a journey that began during my postdoctoral research at the City University of Hong Kong after years in distributed systems. I am deeply grateful to my supervisor, Prof. Xiaohua Jia, for his guidance and encouragement. He taught me that a researcher's greatest asset is the courage to learn anew. His support made this work possible.

Preface

Artificial intelligence (AI) models are trained on large amounts of data, often containing sensitive or regulated information. When certain data must be removed due to privacy requests, legal requirements, or model updates, retraining the entire model from scratch becomes computationally expensive. Machine unlearning addresses this challenge by selectively removing specific training data while preserving model performance, helping AI systems meet legal and ethical requirements efficiently.

This book is organized into two parts. Part I, *Theory of Machine Unlearning*, develops the conceptual and analytical foundations of the field. It begins with the motivations and formal problem definitions of unlearning, followed by an exploration of data influence analysis and its role in understanding model forgetfulness. Subsequent chapters introduce real-time and time-varying influence measurement frameworks that capture dynamic data–model interactions over time. Part II, *Applications of Machine Unlearning*, translates theoretical insights into practical mechanisms for large-scale systems. It presents exact unlearning techniques for large models based on prompt tuning, as well as in-training unlearning methods that enable models to forget data during learning. The final chapter discusses deployment challenges, regulatory implications, and open research directions for advancing machine unlearning in practice.

This book would not exist without the many collaborators, colleagues, and students. Their thoughtful discussions, honest feedback, and careful experiments have shaped this work in important ways, and we are truly thankful for that. We also appreciate the editorial and production teams at Springer Nature for their professionalism and support throughout the publication process.

Kowloon, Hong Kong
January 2026

Jie Xu
Xiaohua Jia

Acknowledgements This book grew out of research I began during my postdoctoral studies at the City University of Hong Kong. It covers both the theory and applications of machine unlearning and represents my first steps into artificial intelligence research. During the preparation of this manuscript, the author used large language model tools (ChatGPT and Claude) for the purposes of improving language, refining text clarity, and assisting with manuscript formatting. The author reviewed and edited all AI-assisted output and takes full responsibility for the content of this book.

I am deeply grateful to my supervisor, Prof. Xiaohua Jia, for his guidance, encouragement, and trust. His way of thinking about problems has shaped my own approach as a researcher, and this book would not exist without his support.

I gratefully acknowledge the Research Grants Council of Hong Kong for providing funding that made portions of this research possible. I also wish to thank my collaborators and lab colleagues. Their feedback and technical contributions have greatly enriched the depth and quality of this book. The ideas we exchanged and the experiments we ran together have made this a much better book. Several chapters draw from papers I published in the *Proceedings of the ACM International Conference on Information and Knowledge Management (CIKM)*, the *Proceedings of the International Conference on Artificial Intelligence and Statistics (AISTATS)*, and *IEEE Transactions on Emerging Topics in Computational Intelligence*. I thank the reviewers and editors at these venues for their careful reading and helpful suggestions, which strengthened the underlying work.

Finally, I am grateful to my parents. Their love, patience, and steady encouragement have sustained me throughout my academic journey. This book is possible because of them.

Competing Interests This monograph includes materials based on our previously published work:

- Jie Xu, Zihan Wu, Cong Wang, and Xiaohua Jia, "Machine Unlearning: Solutions and Challenges," *IEEE Transactions on Emerging Topics in Computational Intelligence*, 2024, pp. 1–19.
- Jie Xu, Zihan Wu, Cong Wang, and Xiaohua Jia, "LiveVal: Real-time and Trajectory-based Data Valuation via Adaptive Reference Points," *Proceedings of the ACM International Conference on Information and Knowledge Management (CIKM)*, 2025.
- Jie Xu, Zihan Wu, Cong Wang, and Xiaohua Jia, "LMEraser: Large Model Unlearning via Adaptive Prompt Tuning," *Proceedings of the International Conference on Artificial Intelligence and Statistics (AISTATS)*, 2025.

These publications have been reused with permission and are properly cited in the relevant chapters.

Ethics Approval This book does not involve studies with human participants or animals.

Contents

Part I Theory of Machine Unlearning

1 Introduction 3
Fundamentals of Machine Learning 3
Key Components of Machine Learning 4
The Process of Machine Learning 9
Limitations of Machine Learning 12
Risks from Private and Copyrighted Data 13
Risks from Malicious and Corrupted Data 14
Risks from Outdated, Biased, and Low-Quality Data 15
Motivations of Machine Unlearning 16
The Machine Unlearning Problem 16
Application Scenarios and Design Principles 19
Machine Unlearning in Wireless Networks 22
Machine Learning Across the Wireless Protocol Stack 22
Unlearning Requirements in Wireless Deployments 23
Design Constraints and Book Roadmap 24
Structure of the Book 26
References 28

2 Background of Machine Unlearning 33
Data Influence Analysis 33
Classical Data Influence Analysis 33
Limitations of Classical Methods 47
Relationships Between Data Influence and Unlearning 52
Exact Machine Unlearning 55
Naïve Retraining and SISA 56
Existing Exact Unlearning Methods 59
Advantages and Limitations 63

Approximate Machine Unlearning 64
Parameter Manipulation Methods 64
Fine-Tuning and Reweighting Methods 69
Advantages and Limitations 74
Evaluation of Machine Unlearning 75
Forgetting Performance 76
Model Utility Performance 81
Benchmarks for Unlearning 84
References 85

3 Real-Time Data Influence Measurement 91
Motivation and Contribution 91
System Model and Problem Statement 92
System Model 93
Problem Statement 93
Trajectory-Based Data Valuation Framework 94
Basic Method with Static Reference Points 95
Real-Time Measurement with Adaptive Reference Points 96
Real-Time Implementation via Dual Queues 99
Theoretical Analysis 101
Assumptions and Fundamental Properties 101
Stability Analysis 102
Fairness Analysis 104
Experimental Evaluation 105
Experimental Setup 105
Comparison Against Baselines 106
Early Detection Capability 108
Robustness Analysis 109
Hyperparameter Sensitivity Analysis 112
Conclusion 113
References 114

4 Time-Varying Data Influence Measurement 117
Motivation and Contribution 117
System Models and Problem Statement 119
System Model 119
Problem Statement 120
Design of Time-Varying Influence Measurement 120
Influence on Model Parameters 121
Influence on Model Behaviors 123
Implementations 124
Standard Implementation 124
Checkpoint-Based Implementation of TVIM 126

Theoretical Analysis 127
Estimation Error Analysis 128
Time-Varying Influence Toolkit 133
Experimental Evaluation 138
Experimental Setup 138
Accuracy of Influence Measurement 139
Influence Dynamics and Similarity Across Training Stages 140
Applications of Time-Varying Influence Measurement 141
Conclusion 142
References 143

Part II Applications of Machine Unlearning

5 Exact Unlearning for Large Models via Prompt Tuning 147
Motivation and Contribution 147
System Model and Problem Statement 149
System Model 149
Problem Statement 150
Design of Exact Unlearning for Large Models 151
Data Partitioning and Pre-training 151
Adaptive Prompt Tuning 152
Prompt-Tuning-Based Unlearning 157
Theoretical Analysis 159
Probabilistic Model for Unlearning Requests 159
Analysis of Affected Clusters 159
Processing Time Analysis and Efficiency Gains 161
Experimental Evaluation 162
Experimental Setup 163
Evaluation of Model Utility 164
Evaluation of Unlearning Efficiency 168
Ablation Studies 170
Conclusion 172
References 172

6 Approximate Unlearning for Large Models via Noise Injection 175
Motivation and Contribution 175
System Model and Problem Statement 177
System Model 177
Problem Statement 179
In-Training Unlearning Framework 180
Influence Decay and Noise Injection 180
Maturity Condition 181
Complete Unlearning Protocol 183
Theoretical Analysis 184
Influence Decay Analysis 185
Certified Unlearning 188

Utility Preservation 189
Complexity Analysis 192
Conclusion 194
References 194

7 Conclusion, Practical Challenges, and Future Directions 197
Conclusions 197
Data Influence as the Theoretical Foundation 197
Unlearning Algorithms as the Practical Framework 199
Practical Challenges in Wireless Network Deployment 200
Resource-Constrained and Heterogeneous Edge Computation 200
Federated Unlearning Under Communication Constraints 201
Verification in Distributed Wireless Systems 202
Future Research Directions 202
Theoretical Extensions 203
Algorithmic Extensions to Emerging Model Paradigms 203
Integration Into Next-Generation Wireless Architectures 204
References 205

Acronyms

ASR	Attack Success Rate
AUC	Area Under the Curve
CCPA	California Consumer Privacy Act
CIFAR	Canadian Institute for Advanced Research (dataset)
CLIP	Contrastive Language–Image Pre-training
CNN	Convolutional Neural Network
COCO	Common Objects in Context (dataset)
CSI	Channel State Information
DL	Deep Learning
DMCA	Digital Millennium Copyright Act
DNN	Deep Neural Network
DP	Differential Privacy
DPDPA	Digital Personal Data Protection Act (India)
DPO	Direct Preference Optimization
EK-FAC	Eigenvalue-corrected Kronecker-Factored Approximate Curvature
EL2N	Error L2 Norm
eMBB	Enhanced Mobile Broadband
EMNIST	Extended Modified National Institute of Standards and Technology (dataset)
ERM	Empirical Risk Minimization
EWC	Elastic Weight Consolidation
FastIF	Fast Influence Functions
FL	Federated Learning
GD	Gradient Descent
GDPR	General Data Protection Regulation
GNN	Graph Neural Network
GPT	Generative Pre-trained Transformer
GPU	Graphics Processing Unit
GraNd	Gradient Normed (data valuation)
GTSRB	German Traffic Sign Recognition Benchmark (dataset)
i.i.d.	Independent and Identically Distributed

IF	Influence Function
IoT	Internet of Things
KL	Kullback–Leibler (divergence)
KNN	K-Nearest Neighbor
KS	Kolmogorov–Smirnov (test)
L-BFGS	Limited-memory Broyden–Fletcher–Goldfarb–Shanno
LGPD	Lei Geral de Proteção de Dados (Brazil)
LiRA	Likelihood Ratio Attack
LiSSA	Linear time Stochastic Second-order Algorithm
LLM	Large Language Model
LOO	Leave-One-Out
LR	Logistic Regression
MAC	Medium Access Control
MEMIT	Mass Editing Memory in a Transformer
MIA	Membership Inference Attack
MIMO	Multiple-Input Multiple-Output
ML	Machine Learning
MMLU	Massive Multitask Language Understanding
mMTC	Massive Machine Type Communications
mmWave	Millimeter Wave
MNIST	Modified National Institute of Standards and Technology (dataset)
MSE	Mean Squared Error
MUSE	Machine Unlearning Six-way Evaluation (benchmark)
NPO	Negative Preference Optimization
OFDM	Orthogonal Frequency-Division Multiplexing
O-RAN	Open Radio Access Network
PIPL	Personal Information Protection Law (China)
QA	Question Answering
RAN	Radio Access Network
RIC	RAN Intelligent Controller
RL	Reinforcement Learning
ROC	Receiver Operating Characteristic
ROME	Rank-One Model Editing
ROUGE	Recall-Oriented Understudy for Gisting Evaluation
SGD	Stochastic Gradient Descent
SISA	Sharded, Isolated, Sliced, and Aggregated
SMOTE	Synthetic Minority Over-sampling Technique
SVHN	Street View House Numbers (dataset)
SWaP	Size, Weight, and Power
TF-IDF	Term Frequency–Inverse Document Frequency
TOFU	Task of Fictitious Unlearning (benchmark)
TracIn	Tracing with Influence
TRAK	Tracing with Randomly-projected Approximations of Kernels
URLLC	Ultra-Reliable Low-Latency Communication

ViT	Vision Transformer
WAIM	Window-Accumulated Influence Measurement
WMDP	Weapons of Mass Destruction Proxy (benchmark)

Part I
Theory of Machine Unlearning

This part lays the theoretical foundation of machine unlearning.

It introduces the machine unlearning problem and its relevance to wireless network deployments (Chap. 1), surveys existing unlearning methods and evaluation frameworks (Chap. 2), and develops two complementary influence measurement theories: a real-time valuation framework that operates during training (Chap. 3) and a framework that captures cumulative influence over arbitrary training intervals (Chap. 4).

Chapter 1
Introduction

Abstract Machine learning systems encode training data influence into model parameters through gradient-based optimization. When training data must be deleted due to privacy regulations, security threats, or data quality degradation, the absence of efficient removal mechanisms becomes a critical limitation. This chapter establishes the foundations for studying machine unlearning. We formalize how data influence propagates through loss computation, gradient updates, and the optimization trajectory, then identify three categories of risk that demand removal: private and copyrighted data subject to legal erasure obligations, malicious and corrupted data requiring security remediation, and outdated or biased data necessitating maintenance. Each category imposes a distinct constraint on the unlearning algorithm: verifiability, removal completeness, and efficiency, respectively. We provide formal definitions of exact and approximate unlearning, map application scenarios to algorithmic pathways, and analyze how wireless network deployments amplify these requirements through edge computation limits, distributed architectures, and real-time service guarantees.

Fundamentals of Machine Learning

Machine learning systems learn models from data, and the properties of the training data fundamentally determine what these models can and cannot do. This section formalizes the five core components of any machine learning system: data, models, loss functions, optimization algorithms, and evaluation frameworks, and characterizes the role that each component plays in determining the behavior of the learned model (Section "Key Components of Machine Learning"). Then, we describe the end-to-end process through which these components interact to produce a deployed model, from data collection through training to deployment and maintenance (Section "The Process of Machine Learning"). Throughout, we emphasize a perspective that is central to the machine unlearning problem: training data exerts a persistent, measurable, and often nonlinearly entangled influence on the model parameters, and understanding this influence at the level of individual components is a prerequisite

J. Xu and X. Jia, *Machine Unlearning*, Wireless Networks,
https://doi.org/10.1007/978-3-032-30978-5_1

for principled data removal. The concepts and notation introduced here are used consistently throughout Chaps. 2 through 7.

Key Components of Machine Learning

Every machine learning system is built from five fundamental components: data ($\mathcal{D}$), model/hypothesis space ($\mathcal{H}$), loss function (ℓ), optimization algorithm (optimizer, $\mathcal{O}$), and evaluation framework ($\mathcal{E}$). Together with an initialization and a set of hyperparameters, these components induce an end-to-end (possibly randomized) training procedure, denoted by $\mathcal{T}$, that maps a training dataset to learned parameters (formalized in Definition 1.7). Throughout this book, we reserve $\mathcal{O}$ for the optimizer (e.g., SGD/Adam) and use $\mathcal{T}$ for the full training procedure, so that all unlearning guarantees in section "Motivations of Machine Unlearning" compare an unlearned output directly against retraining under the same $\mathcal{T}$.

Data.

Data serves as the empirical foundation upon which machine learning systems are constructed [1, 2]. We formalize the data component as follows.

Definition 1.1 (*Dataset*) A dataset is a finite collection $\mathcal{D} = \{z_i\}_{i=1}^{N}$ of N data points drawn from an unknown data-generating distribution P over a data space $\mathcal{Z}$. In supervised learning, each data point takes the form $z_i = (\mathbf{x}_i, y_i)$ where $\mathbf{x}_i \in \mathcal{X} \subseteq \mathbb{R}^d$ is a d-dimensional feature vector and $y_i \in \mathcal{Y}$ is the corresponding label, with $\mathcal{Y} = \{1, \ldots, C\}$ for C-class classification and $\mathcal{Y} = \mathbb{R}$ for regression; in unsupervised learning, $z_i = \mathbf{x}_i$ and no labels are observed.

The standard methodology partitions $\mathcal{D}$ into three disjoint subsets: a *training set* $\mathcal{D}_{\text{train}}$ used for parameter estimation, a *validation set* $\mathcal{D}_{\text{val}}$ used for hyperparameter tuning and model selection, and a *test set* $\mathcal{D}_{\text{test}}$ held out entirely for final performance evaluation [3]. This partition assumes all subsets are drawn i.i.d. from the same distribution P, an assumption that may be violated due to temporal drift, selection bias, or domain shift [4].

Machine unlearning is primarily concerned with training data because training points directly participate in parameter estimation, and their influence becomes encoded in the learned parameters. By design, the test set $\mathcal{D}_{\text{test}}$ should not affect training or model selection; deleting points from $\mathcal{D}_{\text{test}}$ therefore changes only post-hoc evaluation and requires no unlearning algorithm. Validation data $\mathcal{D}_{\text{val}}$ typically does not enter gradient updates, but it can influence the final deployed model indirectly through hyperparameter tuning, early stopping, and model selection. As a result, deleting validation points may necessitate re-running the selection procedure, even though it does not "subtract" influence from a fixed parameter vector in the same sense as unlearning. In contrast, every training data point $z_i \in \mathcal{D}_{\text{train}}$ contributes to the optimization objective during training, and this contribution becomes encoded in the resulting parameters in a manner that generally cannot be disentangled by inspection alone.

Data quality within $\mathcal{D}_{\text{train}}$ directly shapes model behavior: label noise degrades generalization [5], class imbalance biases decision boundaries [6], and insufficient input coverage produces poor out-of-distribution performance [7]. When a training data point must be removed, whether due to regulatory mandates such as GDPR [8] and CCPA [9], the discovery of erroneous or poisoned samples [10, 11], or user-initiated deletion requests, the persistence of its influence becomes a liability. Quantifying and removing this influence while preserving utility from the remaining data is the central technical challenge addressed throughout this book.

Models.

A model defines the hypothesis space, the set of candidate functions, from which the learning algorithm selects an approximation to the unknown input-to-output relationship, and the choice of hypothesis space determines both the expressiveness and the inductive bias of the learner [12, 13].

Definition 1.2 (*Parameterized Hypothesis Space*) A parameterized hypothesis space is a family of functions $\mathcal{H} = \{h_{\boldsymbol{\theta}} : \mathcal{X} \to \mathcal{Y} \mid \boldsymbol{\theta} \in \Theta\}$ indexed by a parameter vector $\boldsymbol{\theta} \in \Theta \subseteq \mathbb{R}^p$, where p denotes the number of learnable parameters. The learning process selects a specific parameter vector $\boldsymbol{\theta}^* \in \Theta$ based on the training data $\mathcal{D}_{\text{train}}$, yielding a concrete hypothesis $h_{\boldsymbol{\theta}^*}$.

Deep neural networks parameterize $h_{\boldsymbol{\theta}}$ as a composition of L layers, $h_{\boldsymbol{\theta}}(\mathbf{x}) = f_L \circ f_{L-1} \circ \cdots \circ f_1(\mathbf{x})$, where each layer f_l applies a parameterized affine transformation followed by a nonlinear activation [1]. This compositional structure enables deep networks to learn hierarchical feature representations and has yielded state-of-the-art performance across computer vision [14], natural language processing [15], and speech recognition [16]. The concept of model capacity formalizes the richness of the function class that $\mathcal{H}$ can represent and is central to understanding generalization behavior: insufficient capacity produces underfitting, while excessive capacity can produce overfitting [1].

From the perspective of machine unlearning, the relationship between model capacity and data memorization has direct implications. High-capacity models, particularly overparameterized deep networks where $p \gg N$, have been shown to interpolate the training data, effectively memorizing each individual training data point [17, 18]. This memorization implies that individual training data points exert stronger and more identifiable influences on the model's behavior, making both the detection and the removal of specific data contributions simultaneously more urgent (due to privacy risks from memorization [19]) and more challenging (due to the complex entanglement of parameters). The interplay between model capacity and unlearning difficulty is a recurring theme in the methods presented in Chaps. 5 and 6.

Loss Functions and Optimization Objectives.

The loss function translates the learning task into a formal optimization problem by quantifying the discrepancy between model predictions and desired outputs [1, 12].

Definition 1.3 (*Loss Function and Empirical Risk*) A loss function $\ell : \mathcal{Y} \times \mathcal{Y} \to \mathbb{R}_{\geq 0}$ maps a prediction-label pair to a non-negative scalar quantifying prediction error. Given a model $h_{\boldsymbol{\theta}}$ and a training dataset $\mathcal{D} = \{(\mathbf{x}_i, y_i)\}_{i=1}^{N}$, the *empirical risk* is the average loss over the training set:

$$\hat{R}(\boldsymbol{\theta}; \mathcal{D}) = \frac{1}{N} \sum_{i=1}^{N} \ell\big(h_{\boldsymbol{\theta}}(\mathbf{x}_i), y_i\big). \tag{1.1}$$

Commonly employed loss functions include the cross-entropy loss $\ell_{\text{CE}}(\hat{\mathbf{p}}, y) = -\log \hat{p}_y$ for classification and the squared loss $\ell_{\text{MSE}}(\hat{y}, y) = (\hat{y} - y)^2$ for regression [2, 3].

The principle of *Empirical Risk Minimization* (ERM) prescribes selecting the model parameters that minimize the empirical risk:

$$\boldsymbol{\theta}^*_{\text{ERM}} = \arg\min_{\boldsymbol{\theta} \in \Theta} \hat{R}(\boldsymbol{\theta}; \mathcal{D}). \tag{1.2}$$

When the hypothesis space $\mathcal{H}$ has high capacity relative to the sample size N, ERM alone is susceptible to overfitting. Regularized risk minimization addresses this by augmenting the objective with a penalty term $\Omega(\boldsymbol{\theta})$ that encodes a preference for simpler models:

$$\boldsymbol{\theta}^*_{\text{reg}} = \arg\min_{\boldsymbol{\theta} \in \Theta} \Big[\hat{R}(\boldsymbol{\theta}; \mathcal{D}) + \lambda\, \Omega(\boldsymbol{\theta}) \Big], \tag{1.3}$$

where $\lambda > 0$ controls the regularization strength. Standard choices include L_2 regularization $\Omega(\boldsymbol{\theta}) = \|\boldsymbol{\theta}\|_2^2$ (weight decay [20]), which encourages small parameter magnitudes, and L_1 regularization $\Omega(\boldsymbol{\theta}) = \|\boldsymbol{\theta}\|_1$ (the Lasso [21]), which promotes parameter sparsity.

The additive structure of the empirical risk in Eq. (1.1) provides a foundational insight for machine unlearning. Each training data point $z_i = (\mathbf{x}_i, y_i)$ contributes a term $\ell(h_{\boldsymbol{\theta}}(\mathbf{x}_i), y_i)/N$ to the aggregate objective. Removing a subset $\mathcal{D}_f \subset \mathcal{D}$ of data points (the *forget set*) from the training data modifies the objective to:

$$\hat{R}(\boldsymbol{\theta}; \mathcal{D} \setminus \mathcal{D}_f) = \frac{1}{N - |\mathcal{D}_f|} \sum_{z_i \in \mathcal{D} \setminus \mathcal{D}_f} \ell\big(h_{\boldsymbol{\theta}}(\mathbf{x}_i), y_i\big). \tag{1.4}$$

In principle, the unlearned model can be obtained by re-solving Eq. (1.2) on $\mathcal{D} \setminus \mathcal{D}_f$, which constitutes *exact unlearning via retraining*. However, this naïve approach incurs the full computational cost of training from scratch, which is prohibitive for large-scale models. The fundamental challenge addressed in Chaps. 5

and 6 is to achieve unlearning that is *equivalent* or *approximately equivalent* to retraining while requiring substantially less computation.

Learning Algorithms and Optimization.

Optimization algorithms constitute the computational mechanism that transforms the declarative specification of a loss function into a concrete model by iteratively updating parameters [1, 22]. The predominant framework in modern machine learning is first-order gradient-based optimization.

Definition 1.4 (*Gradient Descent and Stochastic Gradient Descent*) Given a differentiable empirical risk $\hat{R}(\boldsymbol{\theta}; \mathcal{D})$, *gradient descent* (GD) updates parameters at iteration t as:

$$\boldsymbol{\theta}_{t+1} = \boldsymbol{\theta}_t - \eta_t \nabla_{\boldsymbol{\theta}} \hat{R}(\boldsymbol{\theta}_t; \mathcal{D}), \quad (1.5)$$

where $\eta_t > 0$ is the learning rate at iteration t.

Stochastic gradient descent (SGD) [23] replaces the full gradient with a stochastic approximation computed over a randomly sampled mini-batch $\mathcal{B}_t \subset \mathcal{D}$ of size $|\mathcal{B}_t| = B$:

$$\boldsymbol{\theta}_{t+1} = \boldsymbol{\theta}_t - \frac{\eta_t}{B} \sum_{z_i \in \mathcal{B}_t} \nabla_{\boldsymbol{\theta}} \ell\big(h_{\boldsymbol{\theta}_t}(\mathbf{x}_i),\ y_i\big). \quad (1.6)$$

The stochastic gradient in Eq. (1.6) constitutes an unbiased estimator of the full gradient, with variance decreasing as $O(1/B)$ [22]. Adaptive variants such as momentum methods [24], AdaGrad [25], and Adam [26] improve practical convergence behavior through accumulated gradient statistics and per-parameter learning rate adaptation.

A perspective on optimization that is particularly important for machine unlearning is the interpretation of the training process as a *trajectory* through parameter space [27]. Starting from an initialization $\boldsymbol{\theta}_0$, the optimizer produces a sequence of iterates:

$$\boldsymbol{\theta}_0 \xrightarrow{\mathcal{B}_1} \boldsymbol{\theta}_1 \xrightarrow{\mathcal{B}_2} \boldsymbol{\theta}_2 \xrightarrow{\mathcal{B}_3} \cdots \xrightarrow{\mathcal{B}_T} \boldsymbol{\theta}_T, \quad (1.7)$$

where each transition $\boldsymbol{\theta}_t \rightarrow \boldsymbol{\theta}_{t+1}$ is determined by the mini-batch $\mathcal{B}_t$ drawn at step t. The final model $\boldsymbol{\theta}_T$ is therefore a cumulative function of every mini-batch encountered during training, and a given data point z_i influences $\boldsymbol{\theta}_T$ through every iteration in which $z_i \in \mathcal{B}_t$. Due to the sequential dependence of iterates, the influence of z_i at an early iteration t propagates forward through all subsequent updates, creating a complex chain of dependencies. This trajectory-centric view motivates the data influence measurement frameworks developed in Chaps. 3 and 4, where we formalize and quantify how individual data points shape the parameter trajectory across the entire training process.

Generalization and Evaluation.

The capacity to generalize, to perform accurately on previously unseen data drawn from the same distribution that generated the training set, is the defining objective of machine learning [12]. Generalization is formalized through the population risk.

Definition 1.5 (*Population Risk and Generalization Gap*) The *population risk* (or true risk) of a model h_θ is the expected loss under the data-generating distribution P:

$$R(\boldsymbol{\theta}) = \mathbb{E}_{(\mathbf{x},y)\sim P}\big[\,\ell\big(h_{\boldsymbol{\theta}}(\mathbf{x}),\ y\big)\,\big]. \tag{1.8}$$

The *generalization gap* is defined as the difference between the population risk and the empirical risk:

$$\Delta_{\text{gen}}(\boldsymbol{\theta};\mathcal{D}) = R(\boldsymbol{\theta}) - \hat{R}(\boldsymbol{\theta};\mathcal{D}). \tag{1.9}$$

A learning algorithm generalizes well if Δ_{gen} is small with high probability over the random draw of $\mathcal{D} \sim P^N$.

Since the population risk $R(\boldsymbol{\theta})$ cannot be computed directly (as P is unknown), it is estimated using the held-out test set $\hat{R}(\boldsymbol{\theta};\mathcal{D}_{\text{test}})$. Algorithmic stability provides a framework particularly relevant to machine unlearning: a (possibly randomized) training procedure $\mathcal{T}$ is *uniformly stable* if replacing a single training data point changes the output model's loss by at most β, and Bousquet and Elisseeff [28] established that uniform stability of order β implies a generalization gap bounded by $O(\beta)$. The stability perspective characterizes the sensitivity of the trained model to individual data points, precisely the quantity that unlearning seeks to manipulate.

In the context of machine unlearning, evaluation must address two additional and potentially conflicting objectives beyond standard model performance: *forgetting quality* (the degree to which the target data's influence has been removed) and *model utility preservation* (the degree to which the unlearned model retains its performance on the remaining data). Forgetting quality is assessed using techniques such as membership inference attacks [29], which test whether an adversary can determine whether a specific data point was used in training, and distributional distance measures that compare the unlearned model to a model retrained from scratch without the target data [30].

Summary.

The five components formalized above form an interconnected system in which data plays a foundational and pervasive role. The model parameters $\boldsymbol{\theta}^*$ are determined by minimizing a loss function that is an additive aggregation over training data points (Eq. (1.1)), through an optimization trajectory (Eq. (1.7)) in which each training data point influences every subsequent parameter update. The influence of any single training data point z_i is propagated through the loss, encoded in the trajectory, and embedded in the final parameters in a manner that is generally nonlinear and non-decomposable. Removing this influence, the core task of machine unlearning, therefore, requires interventions that account for the interactions across all five components.

The Process of Machine Learning

The standard machine learning pipeline orchestrates the components formalized in section "Key Components of Machine Learning" through a sequence of stages that collectively produce, evaluate, and maintain a deployed model. We decompose this process into six stages: problem formulation, data acquisition and preprocessing, model training, model evaluation and selection, deployment and serving, and post-deployment monitoring and adaptation.

Stage 1: Problem Formulation.

The machine learning process begins with the formulation of a well-defined learning task, which specifies the input space $\mathcal{X}$, the output space $\mathcal{Y}$, and the performance criterion by which success is measured [12]. From the perspective of unlearning, the problem formulation stage determines the *scope* of potential unlearning requests: a classification model trained on user data may receive sample-level deletion requests (remove the influence of specific users), class-level requests (remove the ability to classify a particular category), or feature-level requests (remove dependence on a sensitive attribute) [31, 32]. Anticipating the types of unlearning requests at the formulation stage can inform architectural decisions that facilitate more efficient removal, as we discuss in Chap. 5.

Stage 2: Data Acquisition and Preprocessing.

Data acquisition assembles the training corpus $\mathcal{D}$ from one or more sources and applies preprocessing transformations to produce a representation suitable for model training [1, 3]. We formalize the preprocessing pipeline as a sequence of deterministic transformations applied to raw data.

Definition 1.6 (*Data Preprocessing Pipeline*) A preprocessing pipeline is a composition of K transformations $\phi = \phi_K \circ \phi_{K-1} \circ \cdots \circ \phi_1$, where each $\phi_k : \mathcal{Z}^{(k-1)} \to \mathcal{Z}^{(k)}$ maps from an intermediate representation space $\mathcal{Z}^{(k-1)}$ to the next. The raw dataset $\mathcal{D}^{(0)} = \{z_i^{(0)}\}_{i=1}^{N_0}$ is transformed into the training-ready dataset $\mathcal{D} = \{z_i\}_{i=1}^{N}$ via $\mathcal{D} = \phi(\mathcal{D}^{(0)})$, where N may differ from N_0 due to filtering or augmentation operations.

Stage 3: Model Training.

Model training is the stage at which data influence is most directly and extensively inscribed into the model parameters [1, 22]. The training process operationalizes the ERM principle (Eq. (1.2)) or its regularized variant (Eq. (1.3)) by executing an optimization algorithm (Definition 1.4) over the training set $\mathcal{D}_{\text{train}}$ for a prescribed number of iterations or until a convergence criterion is satisfied.

Definition 1.7 (*Training Procedure*) A training procedure is a tuple

$$\mathcal{T} = (\mathcal{H},\ \ell,\ \mathcal{O},\ \boldsymbol{\theta}_0,\ \boldsymbol{\psi}), \tag{1.10}$$

consisting of a hypothesis space $\mathcal{H}$ (Definition 1.2), a loss function ℓ (Definition 1.3), an optimization algorithm $\mathcal{O}$ (e.g., SGD per Definition 1.4), an initialization $\boldsymbol{\theta}_0 \in \Theta$, and a vector of hyperparameters

$$\boldsymbol{\psi} = (\eta,\ B,\ \lambda,\ T,\ \ldots), \tag{1.11}$$

where η denotes the learning rate, B the mini-batch size, λ the regularization strength (Eq. (1.3)), and T the total number of parameter updates.

Executing $\mathcal{T}$ on a training set $\mathcal{D}_{\text{train}}$ produces trained parameters $\boldsymbol{\theta}_T$; we denote the corresponding learned hypothesis by $h_{\boldsymbol{\theta}_T}$:

$$\boldsymbol{\theta}_T = \mathcal{T}(\mathcal{D}_{\text{train}}), \qquad h_{\boldsymbol{\theta}_T} \in \mathcal{H}. \tag{1.12}$$

In practice, $\mathcal{T}(\mathcal{D}_{\text{train}})$ is obtained by running the iterative update rule specified by $\mathcal{O}$ (e.g., Eq. (1.6) for SGD or an adaptive variant thereof) for $t = 0, 1, \ldots, T-1$.

The training process proceeds in *epochs*, where one epoch corresponds to a complete pass over the training set $\mathcal{D}_{\text{train}}$. Training for E epochs involves $T = E \cdot \lceil N/B \rceil$ total parameter updates, each of which is shaped by a specific subset of the training data. The total computational cost of training scales as $O(E \cdot N \cdot p)$, where p is the number of model parameters [1].

Several phenomena that arise during training have direct relevance to unlearning. *Training stochasticity*, arising from random mini-batch sampling and random initialization, implies that the same dataset can produce different models across independent runs. This variability complicates the definition of exact unlearning, since the "gold standard" retrained model $\boldsymbol{\theta}^*_{\text{retrain}}$ is itself a random variable rather than a deterministic quantity. *Gradient accumulation* over multiple epochs means that frequently encountered data points exert disproportionate cumulative influence on the parameter trajectory (Eq. (1.7)), creating a non-uniform distribution of data influence that the measurement frameworks of Chaps. 3 and 4 are designed to capture.

Stage 4: Model Evaluation and Selection.

Model evaluation and selection identify the best-performing model configuration from a set of candidates by measuring performance on the held-out validation set $\mathcal{D}_{\text{val}}$ [3, 33]. The evaluation stage interacts with unlearning in that the validation set may itself contain data points subject to deletion requests: removing a targeted data point $z_j \in \mathcal{D}_{\text{val}}$ may alter the model selection criterion and, consequently, the selected configuration [30]. Additionally, when the deployment context involves potential deletion requests, the evaluation protocol must incorporate unlearning-specific criteria beyond standard predictive performance, including forgetting quality and utility preservation. The formal evaluation framework for these criteria is developed in Chap. 2.

Stage 5: Deployment and Serving.

Deployment transitions the trained model from a development artifact to a production system, introducing operational constraints that fundamentally shape the feasibility and urgency of unlearning [34, 35].

Definition 1.8 (*Deployed Model and Inference*) A deployed model is a fixed function $h_{\boldsymbol{\theta}_T} : \mathcal{X} \to \mathcal{Y}$ parameterized by the trained weights $\boldsymbol{\theta}_T$, embedded in a serving system that accepts input queries and returns predictions. The model operates in *inference mode*: parameters are not updated, and no training data is accessed during serving.

The standard deployment paradigm treats the trained model as a *static artifact*: the parameter vector $\boldsymbol{\theta}_T$ is frozen after training and serves predictions indefinitely. This "train once, serve indefinitely" paradigm rests on the assumption that the data-generating distribution P remains stationary and that no modifications to the training data are required after the model is finalized. The static model assumption creates a fundamental tension with unlearning requirements. Once $\boldsymbol{\theta}_T$ is deployed, it encodes the influence of every training data point, and this influence persists for the entire lifetime of the deployment. If a user exercises a deletion right under GDPR Article 17 [8] or a data point is discovered to be poisoned [10], the naïve solution of full retraining on $\mathcal{D} \setminus \mathcal{D}_f$ incurs the original training cost $O(E \cdot N \cdot p)$ plus the operational overhead of model replacement. For large-scale systems where training requires weeks of computation on hundreds of accelerators [36, 37], this cost renders naïve retraining impractical for all but the rarest unlearning requests. The development of efficient unlearning algorithms (Chaps. 5 and 6) is motivated precisely by the need to bridge this gap between the urgency of data removal and the prohibitive cost of retraining.

Stage 6: Post-Deployment Monitoring and Adaptation.

Post-deployment monitoring detects degradation in model performance caused by changes in the data-generating distribution, and adaptation mechanisms update the model to restore accuracy under the evolving distribution [38, 39].

Definition 1.9 (*Concept Drift*) Let P_t denote the data-generating distribution at time t. Concept drift occurs when $P_t \neq P_{t'}$ for some $t \neq t'$. The drift is classified as: *virtual drift* if only the marginal $P_t(\mathbf{x})$ changes while the posterior $P_t(y \mid \mathbf{x})$ remains fixed; *real drift* if $P_t(y \mid \mathbf{x})$ changes; or *both* if both components shift simultaneously [39].

Online learning updates the model parameters continuously as new data arrives:

$$\boldsymbol{\theta}_{t+1} = \boldsymbol{\theta}_t - \eta_t \nabla_{\boldsymbol{\theta}} \ell\big(h_{\boldsymbol{\theta}_t}(\mathbf{x}_{t+1}),\ y_{t+1}\big), \tag{1.13}$$

which incrementally incorporates the influence of new data into the model. *Continual learning* extends this paradigm to sequential tasks, where the model must learn new tasks while retaining performance on previous ones [40]. The relationship between continual learning and machine unlearning is both complementary and contrastive: continual learning addresses the problem of selectively *retaining* knowledge while incorporating new information, whereas machine unlearning addresses the converse problem of selectively *removing* specific knowledge while preserving the remainder. Despite this conceptual symmetry, machine unlearning demands verifiable removal that can withstand adversarial auditing [29, 30], a requirement that has no counterpart in continual learning.

A critical observation is that the existing model maintenance paradigms all provide mechanisms for *adding* new knowledge to a model, but none provide a principled mechanism for *removing* previously learned knowledge. Online learning (Eq. (1.13)) can only move the parameters forward along the gradient of new data; it cannot invert the influence of past data. This architectural asymmetry represents a fundamental limitation of the standard machine learning pipeline that machine unlearning addresses.

Summary and Transition.

The six-stage pipeline reveals a system: data influence enters through acquisition (Stage 2), is inscribed into parameters during training (Stage 3), persists unchanged throughout deployment (Stage 5), and cannot be subtracted by post-deployment adaptation (Stage 6). Every training data point's contribution to the deployed model is permanent by default, a property that conflicts with the growing legal, ethical, and security demands for selective data removal analyzed in section "Limitations of Machine Learning".

Limitations of Machine Learning

Trained models inherit the characteristics of their training data, including harmful ones. Section "Fundamentals of Machine Learning" established that training data exerts a persistent, nonlinearly entangled influence on model parameters with no mechanism for targeted removal. This section examines how this persistence creates distinct categories of risk.

We organize the analysis around three categories corresponding to fundamentally different data pathologies. Section "Risks from Private and Copyrighted Data" examines legitimate, high-quality data whose continued model influence conflicts with privacy rights or intellectual property rights. Section "Risks from Malicious and Corrupted Data" examines adversarially injected data designed to corrupt model behavior. Section "Risks from Outdated, Biased, and Low-Quality Data" examines data that has become outdated, biased, or defective after deployment. Each category entails distinct threat models, urgency profiles, and unlearning requirements, formalized in section "Motivations of Machine Unlearning".

Risks from Private and Copyrighted Data

Trained models can leak private information and memorize copyrighted content. In both cases, the data was legitimately collected and benefits model performance; the problem is that third parties have legal rights to its removal.

Privacy Leakage.

The persistence of data point influence creates exploitable channels through which adversaries can infer properties of training data [19, 29, 41].

Definition 1.10 (*Membership Inference Attack*) Let h_{θ^*} be a model trained on $\mathcal{D}_{\text{train}}$. A *membership inference attack* (MIA) is a (possibly randomized) test $\mathcal{M}: \mathcal{Z} \times \mathcal{H} \to \{0, 1\}$ that outputs 1 if it infers that a target point z was in $\mathcal{D}_{\text{train}}$, and 0 otherwise. The attack is considered successful if it performs better than random guessing under the assumed membership prior (e.g., achieves AUC > 0.5, or achieves accuracy exceeding the base-rate classifier).

The *Likelihood Ratio Attack* (LiRA) [42] estimates membership by computing:

$$\Lambda(z) = \frac{p\big(\ell(h_{\theta^*}(\mathbf{x}), y) \mid z \in \mathcal{D}_{\text{train}}\big)}{p\big(\ell(h_{\theta^*}(\mathbf{x}), y) \mid z \notin \mathcal{D}_{\text{train}}\big)}. \tag{1.14}$$

Definition 1.11 (*Model Inversion Attack*) Given a trained model h_{θ^*} and target label y^*, a *model inversion attack* reconstructs training data by solving:

$$\hat{\mathbf{x}} = \arg\min_{\mathbf{x} \in \mathcal{X}} \mathcal{L}_{\text{inv}}(h_{\theta^*}, \mathbf{x}, y^*) + \gamma\, \mathcal{R}(\mathbf{x}), \tag{1.15}$$

where $\mathcal{L}_{\text{inv}}$ measures compatibility with the model's behavior for y^* and $\mathcal{R}(\mathbf{x})$ encodes domain priors [41, 43].

Beyond these attacks, *training data extraction* recovers verbatim content from large language models [19, 44]. Feldman [45] established that such memorization is *necessary* for optimal generalization on long-tailed distributions, implying a fundamental tension between memorization and privacy.

Legal Frameworks.

Privacy vulnerabilities have been translated into binding legal obligations. The GDPR's Article 17 establishes a *right to erasure* requiring deletion "without undue delay" [8]. The technical question is whether "erasure" requires removing data influence from model parameters, not just raw data from storage [46]. Similar rights exist under CCPA [9], Brazil's LGPD [47], China's PIPL [48], and India's DPDPA [49]. For intellectual property, large models trained on copyrighted content can reproduce it verbatim [19], generating deletion obligations through DMCA takedowns.

Summary.

Private and copyrighted data risks share a distinctive profile: high-quality data subject to *external rights* claims with legally mandated response timeframes (e.g., GDPR generally requires responding *within one month*, with limited extensions in complex cases). The model must withstand adversarial privacy auditing (MIA, LiRA) and cease reproducing protected content. These characteristics make *verifiability* the binding constraint.

Risks from Malicious and Corrupted Data

Adversaries can inject crafted data points into training to corrupt model behavior. Unlike the previous category, this data *should never have entered the pipeline*: it is adversarially designed content, not legitimate data subject to rights claims.
Data Poisoning.

Definition 1.12 (*Data Poisoning Attack*) A *data poisoning attack* injects poisoned points $\mathcal{D}_p$ into training, producing a contaminated model on $\mathcal{D}_{\text{train}} \cup \mathcal{D}_p$. The objective is either *indiscriminate* (maximizing population risk) or *targeted* (causing specific misclassifications while maintaining low overall error) [10, 50].

The *clean-label poisoning* attack [11] manipulates feature representations without altering labels:

$$\mathbf{x}_j^p = \arg\min_{\mathbf{x}} \left\| f_L(\mathbf{x}) - f_L(\mathbf{x}^*) \right\|_2^2 + \beta \left\| \mathbf{x} - \mathbf{x}_{\text{base}} \right\|_2^2, \tag{1.16}$$

where $f_L(\cdot)$ is the penultimate-layer representation, $\mathbf{x}^*$ is the target, and $\mathbf{x}_{\text{base}}$ is a base point from the target class.

Backdoor Attacks.

Definition 1.13 (*Backdoor Attack*) A *backdoor attack* injects points $z_j^b = (\mathbf{x}_j \oplus \boldsymbol{\delta}, y_{\text{target}})$, where $\boldsymbol{\delta}$ is a trigger pattern. The backdoored model satisfies: (i) *utility preservation* on clean inputs, and (ii) *attack effectiveness*, outputting y_{target} when the trigger is present [51, 52].

As few as 1% trigger-stamped samples suffice for near-perfect attack success while maintaining clean accuracy [51]. Advanced variants use invisible perturbations [53] and sample-specific triggers [54].

Summary.

Malicious data risks involve adversarially injected content, a problem of *data integrity*. Removal is triggered by internal security detection with the highest urgency.

The forget set is inherently *uncertain*: detection produces estimates that may miss true malicious points. These characteristics make *removal completeness* the binding constraint.

Risks from Outdated, Biased, and Low-Quality Data

Training data can degrade over time through distributional shift, discovered biases, or identified quality defects. Unlike malicious data, these points were not adversarially injected; unlike private data, they are not subject to external rights claims. The problem is that their continued influence has become harmful.
Concept Drift.
When the data-generating distribution shifts ($P_t \neq P_{t'}$, Definition 1.9), historical training samples become *actively harmful*: models encode decision boundaries misaligned with the current distribution [38, 39]. Online learning and continual learning can *add* new knowledge but cannot *remove* outdated knowledge. Machine unlearning provides this complementary capability.

Data Quality.

Label noise, duplicates, and corrupted features are pervasive in real-world pipelines [5]. For large-scale datasets, error rates of 1–5% are common [55]. Once identified via influence scoring [56] or confident learning [55], unlearning enables selective removal without discarding knowledge from the remaining data.

Fairness and Bias.

Definition 1.14 (*Group Fairness Criteria*) Let $A \in \{0, 1\}$ denote a sensitive attribute, Y the true label, and $\hat{Y}$ the prediction. *Demographic parity* requires $\Pr[\hat{Y} = 1 \mid A = 0] = \Pr[\hat{Y} = 1 \mid A = 1]$. *Equalized odds* [57] requires equal true/false positive rates across groups. *Predictive parity* requires equal positive predictive values across groups.

Training data bias manifests through historical bias, representation bias, measurement bias, and aggregation bias [58, 59]. When biased sources are identified post-deployment [60], unlearning offers targeted correction.

Summary.

Outdated, biased, and low-quality data risks involve *gradual degradation* discovered through internal audits. The binding constraint is *efficiency*: corrections are high-frequency maintenance operations. Utility preservation is critical since the goal is targeted improvement.

Motivations of Machine Unlearning

Machine unlearning is the task of efficiently removing the influence of specified training data from an already-trained model. Section "Limitations of Machine Learning" established three categories of data-related risk that share a common root cause: the absence of an efficient mechanism for selectively removing data influence from trained models. This section formalizes *what* must be removed and *how* the algorithmic design priorities differ across the three data categories.

We proceed in two stages. First, we provide a formal definition of the machine unlearning problem, distinguishing between exact and approximate unlearning as the two principal solution paradigms (Section "The Machine Unlearning Problem"). Second, we analyze three application scenarios in direct correspondence with the three risk categories of section "Limitations of Machine Learning" (Section "Application Scenarios and Design Principles").

The Machine Unlearning Problem

Machine unlearning asks whether a model can be efficiently updated to behave as though certain training data had never been seen. This subsection provides the formal mathematical foundation for this question, building upon the notation established in section "Key Components of Machine Learning". We state the unlearning problem in its most general form and then distinguish between exact and approximate unlearning.

Core Definitions

Definition 1.15 (*Forget Set and Retain Set*) Given a training dataset $\mathcal{D} = \{z_i\}_{i=1}^N$ and an unlearning request specifying a subset of data to be removed, the *forget set* $\mathcal{D}_f \subset \mathcal{D}$ is the collection of data points whose influence is to be eliminated from the model. The *retain set* is the complement $\mathcal{D}_r = \mathcal{D} \setminus \mathcal{D}_f$, consisting of data points whose influence should be preserved. The partitioning satisfies $\mathcal{D}_f \cup \mathcal{D}_r = \mathcal{D}$ and $\mathcal{D}_f \cap \mathcal{D}_r = \emptyset$.

The forget set $\mathcal{D}_f$ may be specified at different granularity levels depending on the application scenario: *data-point-level* requests target specific individual training data points; *class-level* requests target all data points from specified classes; *user-level* requests target all data points contributed by specified data owners, requiring provenance metadata (Section "The Process of Machine Learning"); and *feature-level* requests target the model's dependence on a specified subset of input dimensions, the most technically challenging granularity since the targeted features are present across the entire dataset [61]. The appropriate granularity is determined by the application scenario, as we analyze in section "Application Scenarios and Design Principles".

Definition 1.16 (*Machine Unlearning Problem*) Let $\mathcal{D} = \{z_i\}_{i=1}^N$ denote the *training dataset* used to fit a released model, and let $\mathcal{T}$ denote the (possibly randomized) training procedure (Definition 1.7). Training on $\mathcal{D}$ produces parameters $\boldsymbol{\theta}^* = \mathcal{T}(\mathcal{D})$.

Given an unlearning request specifying a forget set $\mathcal{D}_f \subset \mathcal{D}$ with retain set $\mathcal{D}_r = \mathcal{D} \setminus \mathcal{D}_f$, the *machine unlearning problem* is to compute updated parameters $\boldsymbol{\theta}^u$ satisfying two requirements.

Removal guarantee: the distribution of $\boldsymbol{\theta}^u$ should satisfy a formal indistinguishability criterion with respect to $\boldsymbol{\theta}^{\setminus f} = \mathcal{T}(\mathcal{D}_r)$, i.e., the parameters obtained by retraining from scratch on the retain set under the same training procedure $\mathcal{T}$.

Efficiency constraint: the computational cost of producing $\boldsymbol{\theta}^u$ must be strictly less than rerunning the full training procedure on the retain set, i.e.,

$$C(\mathcal{U}) < C(\mathcal{T}(\mathcal{D}_r)). \tag{1.17}$$

An operator $\mathcal{U} : \Theta \times \mathcal{Z}^* \times 2^{\mathcal{Z}} \to \Theta$ that maps $(\boldsymbol{\theta}^*, \mathcal{D}, \mathcal{D}_f) \mapsto \boldsymbol{\theta}^u$ is called an *unlearning algorithm*.

The removal guarantee admits multiple instantiations. The strongest criterion requires *distributional equivalence* (exact unlearning). A relaxed criterion permits *approximate indistinguishability* with bounded divergence (approximate unlearning). We formalize both paradigms below.

Definition 1.17 (*(ε, δ)-Certified Unlearning*) Fix a training procedure $\mathcal{T}$ (Definition 1.7), a training dataset $\mathcal{D}$, and a forget set $\mathcal{D}_f \subset \mathcal{D}$ with retain set $\mathcal{D}_r = \mathcal{D} \setminus \mathcal{D}_f$. Let $\boldsymbol{\theta}^* = \mathcal{T}(\mathcal{D})$ be the released parameters, let $\boldsymbol{\theta}^{\setminus f} = \mathcal{T}(\mathcal{D}_r)$ be the parameters obtained by retraining from scratch on the retain set under the same $\mathcal{T}$, and let $\boldsymbol{\theta}^u = \mathcal{U}(\boldsymbol{\theta}^*, \mathcal{D}, \mathcal{D}_f)$ be the output of an unlearning algorithm.

We say $\mathcal{U}$ achieves (ε, δ)-*certified unlearning* for this request if, for any measurable set $W \subseteq \mathbb{R}^p$,

$$\Pr\big[\boldsymbol{\theta}^u \in W\big] \leq e^{\varepsilon} \Pr\big[\boldsymbol{\theta}^{\setminus f} \in W\big] + \delta, \tag{1.18}$$
$$\Pr\big[\boldsymbol{\theta}^{\setminus f} \in W\big] \leq e^{\varepsilon} \Pr\big[\boldsymbol{\theta}^u \in W\big] + \delta. \tag{1.19}$$

Exact Unlearning

Exact unlearning algorithms produce updated parameters whose distribution is identical to that of a model retrained from scratch on the retain set, providing the strongest possible removal guarantee [46, 62].

Definition 1.18 (*Exact Unlearning*) An unlearning algorithm $\mathcal{U}$ achieves *exact unlearning* (with respect to a fixed training procedure $\mathcal{T}$) if, for all datasets $\mathcal{D}$, all forget sets $\mathcal{D}_f \subset \mathcal{D}$, and all measurable subsets $S \subseteq \Theta$,

$$\Pr\big[\mathcal{U}(\boldsymbol{\theta}^*, \mathcal{D}, \mathcal{D}_f) \in S\big] = \Pr\big[\mathcal{T}(\mathcal{D} \setminus \mathcal{D}_f) \in S\big], \tag{1.20}$$

where $\boldsymbol{\theta}^* = \mathcal{T}(\mathcal{D})$, and the probabilities are taken over the randomness in $\mathcal{U}$ and $\mathcal{T}$, respectively.

Naïve retraining trivially satisfies Eq. (1.20) but violates the efficiency constraint of Definition 1.16. The central challenge is to achieve distributional equivalence with substantial computational savings. The SISA framework [62] addresses this through data partitioning, training independent constituent models on disjoint shards, and retraining only the affected shard upon deletion. The exact unlearning methods developed in Chap. 5 extend these approaches to large-scale vision models through prompt-tuning-based strategies that localize the data-dependent parameters to small, independently retrainable modules.

Approximate Unlearning

Approximate unlearning algorithms relax the distributional equivalence requirement to achieve greater computational efficiency [63–65]. The relaxation is formalized using the same mathematical structure as differential privacy, which bounds the distributional distance between algorithm outputs under neighboring inputs.

Definition 1.19 (*Differential Privacy* [66]) A randomized mechanism $\mathcal{M} : \mathcal{Z}^N \to \Theta$ satisfies (ε, δ)-*differential privacy* if for all neighboring datasets $\mathcal{D}$ and $\mathcal{D}'$ differing in at most one data point, and for all measurable subsets $S \subseteq \Theta$,

$$\Pr[\mathcal{M}(\mathcal{D}) \in S] \leq e^{\varepsilon} \cdot \Pr[\mathcal{M}(\mathcal{D}') \in S] + \delta. \tag{1.21}$$

Adapting this structure to the unlearning setting yields the following definition.

Definition 1.20 (*Approximate Unlearning*) An unlearning algorithm $\mathcal{U}$ achieves (ε, δ)-*approximate unlearning* (with respect to a fixed training procedure $\mathcal{T}$) if for all training datasets $\mathcal{D}$, all forget sets $\mathcal{D}_f \subset \mathcal{D}$, and all measurable subsets $S \subseteq \Theta$, letting $\boldsymbol{\theta}^* = \mathcal{T}(\mathcal{D})$, $\boldsymbol{\theta}^{\setminus f} = \mathcal{T}(\mathcal{D} \setminus \mathcal{D}_f)$, and $\boldsymbol{\theta}^u = \mathcal{U}(\boldsymbol{\theta}^*, \mathcal{D}, \mathcal{D}_f)$, we have

$$\Pr\left[\boldsymbol{\theta}^u \in S\right] \leq e^{\varepsilon} \cdot \Pr\left[\boldsymbol{\theta}^{\setminus f} \in S\right] + \delta, \tag{1.22}$$

$$\Pr\left[\boldsymbol{\theta}^{\setminus f} \in S\right] \leq e^{\varepsilon} \cdot \Pr\left[\boldsymbol{\theta}^u \in S\right] + \delta. \tag{1.23}$$

When $\varepsilon = 0$ and $\delta = 0$, the definition reduces to exact unlearning (Definition 1.18).

Although differential privacy and approximate unlearning share the same mathematical structure for measuring distributional closeness, they address distinct objectives: differential privacy bounds the information leakage about any single training data point during training, whereas approximate unlearning bounds the distributional distance between the unlearned model and the retrained model after a specific forget set is removed. Guo et al. [63] introduced the certified removal framework, applying a Newton step correction:

$$\boldsymbol{\theta}^u = \boldsymbol{\theta}^* + \frac{1}{N - |\mathcal{D}_f|} \mathbf{H}_{\mathcal{D}_r}^{-1} \sum_{z_i \in \mathcal{D}_f} \nabla_{\boldsymbol{\theta}} \ell\big(h_{\boldsymbol{\theta}^*}(\mathbf{x}_i),\, y_i\big), \tag{1.24}$$

where $\mathbf{H}_{\mathcal{D}_r} = \nabla_{\boldsymbol{\theta}}^2 \hat{R}(\boldsymbol{\theta}^*; \mathcal{D}_r)$ is the Hessian of the empirical risk. This update is computationally efficient for convex losses and small forget sets but requires $O(p^2)$ storage and $O(p^3)$ computation for the Hessian, which is prohibitive for deep networks where p may exceed 10^9. The approximate unlearning methods developed in Chap. 6 take a fundamentally different approach: rather than explicitly computing and subtracting influence via Hessian inversion, they exploit the natural decay of data influence under weight decay and inject calibrated Gaussian noise to mask the residual, achieving (ε, δ)-certified unlearning at $O(p)$ per-request cost.

Application Scenarios and Design Principles

Different types of problematic data require different unlearning strategies. The formal definitions of section "The Machine Unlearning Problem" provide a unified mathematical framework, but the *design priorities* vary substantially depending on the motivating risk category. This subsection analyzes three application scenarios in direct correspondence with the three risk categories of section "Limitations of Machine Learning", specifying for each the operational context, binding design constraints, and algorithmic pathway through the methods developed in this book.

Unlearning for Private and Copyrighted Data

This scenario responds to external rights claims from data subjects exercising privacy rights under GDPR Article 17 [8], CCPA Sect. 1798.105 [9], or analogous legislation, and from content owners asserting intellectual property rights. The data targeted for removal was legitimately collected and is of high quality; the problem is entirely one of nonnegotiable external demands for influence erasure.

The request granularity is predominantly user-level or data-point-level, and legally mandated response timeframes (GDPR's "without undue delay" is typically operationalized as responding within one month, with limited extensions in complex cases [8]) impose hard deadlines. Among the four requirements (verifiability, removal completeness, efficiency, utility preservation), this scenario imposes the strongest demands on *verifiability*: the data subject, a regulatory authority, or a court may audit the completeness of removal using adversarial techniques such as membership inference attacks (Definition 1.10) and the LiRA test (Eq. (1.14)). For copyright claims, evaluation additionally requires demonstrating that the model can no longer reproduce or closely paraphrase the protected content [67, 68].

These requirements favor *exact unlearning* (Definition 1.18) as the preferred solution paradigm, since distributional equivalence provides the strongest verifiable guarantee. The exact unlearning methods of Chap. 5 are directly applicable. When the volume of deletion requests is high, approximate unlearning (Definition 1.20) with stringent (ε, δ) budgets offers a more efficient alternative. The data influence measurement frameworks of Chaps. 3 and 4 support this scenario by quantifying each data point's contribution to the model parameters, and the noise-injection-based mechanisms of Chap. 6 can handle streams of consent withdrawal requests during ongoing training.

Unlearning for Malicious and Corrupted Data

This scenario responds to the detection of adversarial data that should never have entered the training pipeline: poisoned samples (Definition 1.12), backdoored samples (Definition 1.13), or other adversarially corrupted data. The problem is one of data integrity rather than data rights.

The request granularity is predominantly data-point-level, triggered by internal security detection with the highest urgency among all scenarios. A defining characteristic is the inherent *uncertainty* in the forget set: detection algorithms produce estimates $\hat{\mathcal{D}}_f$ that may suffer from false negatives or false positives. The binding constraint is *removal completeness*: incomplete removal of a backdoor's influence may leave residual trigger sensitivity. Formally, the unlearning must ensure:

$$\Pr\left[h_{\theta^u}(\mathbf{x} \oplus \boldsymbol{\delta}) = y_{\text{target}}\right] \leq \Pr\left[h_{\theta \setminus f}(\mathbf{x} \oplus \boldsymbol{\delta}) = y_{\text{target}}\right] + \epsilon_{\text{sec}}, \tag{1.25}$$

where ϵ_{sec} is a security tolerance and $h_{\theta \setminus f}$ is the retrained clean model.

The influence measurement tools of Chaps. 3 and 4 enable the identification of candidate malicious data points through influence scoring. The approximate unlearning methods of Chap. 6 provide the most appropriate algorithmic pathway, as the emphasis on rapid response favors efficient approximate methods over the more costly exact methods, provided that removal completeness is sufficient to neutralize the attack.

Unlearning for Outdated, Biased, and Low-Quality Data

This scenario responds to the discovery that training data has degraded in validity, whether through distributional shift, fairness audits, or quality assessments. The data was not maliciously injected and is not subject to external rights claims; the problem is one of data quality driven by the model operator's own standards.

The request granularity is the most diverse: data-point-level for mislabeled records, class-level for obsolete categories, and feature-level for the removal of dependence on sensitive attributes. The trigger is internal, of lower urgency but higher frequency than the privacy or security triggers. The binding constraint is *computational efficiency*: a dataset of $N = 10^6$ with a 1% label error rate may require the removal of 10^4 data points, making algorithms with per-request cost approach-

Table 1.1 Comparative design profiles for the three machine unlearning scenarios

Design dimension	Private & copyrighted	Malicious & corrupted	Outdated, biased & low-quality
Forget set certainty	High (user specifies)	Low (detection-dependent)	Medium (audit-dependent)
Typical granularity	User-/Data-point-level	Data-point-level	Data-point-/Class-/Feature-level
Trigger source	External (legal/rights)	Internal (security incident)	Internal (audit/monitoring)
Urgency	Moderate (one-month deadline; limited extensions)	Highest (active compromise)	Low but high frequency
Binding constraint	Verifiability	Removal completeness	Efficiency
Secondary constraint	Removal completeness	Efficiency	Utility preservation
Preferred paradigm	Exact unlearning	Approximate (tight budget)	Approximate (relaxed budget)
Acceptable trade-off	Moderate efficiency cost	Slight utility degradation	Relaxed removal guarantee

ing retraining operationally infeasible. Utility preservation is simultaneously critical because the goal is targeted correction rather than wholesale influence erasure.

The bias correction use case presents a distinctive challenge: the goal is not to remove the influence of specific data points entirely but to adjust the model's behavior to satisfy a fairness criterion (Definition 1.14) while preserving utility. This requires targeting the *discriminatory component* of a data point's influence rather than its entire contribution. The data influence measurement frameworks of Chaps. 3 and 4 enable such decomposition, while the noise-injection-based unlearning of Chap. 6 minimizes amortized per-request cost by integrating removal directly into the training process.

Comparative Summary.

Table 1.1 summarizes the key distinctions across the three scenarios. A common thread is that the data influence measurement tools developed in Chaps. 3 and 4 play a dual role: they support both the *identification* of problematic data and the *guidance* of the unlearning algorithm. The exact and approximate unlearning methods of Chaps. 5 and 6 then execute the removal, with the choice of paradigm determined by the binding constraint.

Machine Unlearning in Wireless Networks

Modern wireless communication systems increasingly rely on machine learning as a core operational component rather than an auxiliary optimization tool, and this pervasive integration creates unlearning requirements that are both urgent and technically distinctive. This section establishes the connection between the theoretical framework developed in sections "Fundamentals of Machine Learning" through "Motivations of Machine Unlearning" and the operational realities of wireless network deployments. We survey the breadth of machine learning's integration into the wireless protocol stack (Section "Machine Learning Across the Wireless Protocol Stack"), analyze how the three categories of data risk manifest under wireless-specific conditions (Section "Unlearning Requirements in Wireless Deployments"), and examine the resource and architectural constraints that shape unlearning algorithm design in this domain (Section "Design Constraints and Book Roadmap").

Machine Learning Across the Wireless Protocol Stack

Data-driven models have penetrated every layer of the wireless protocol stack. At the physical layer, deep neural networks perform channel estimation [69], signal detection in OFDM systems [70], and hybrid beamforming optimization for mmWave massive MIMO arrays [71]. At the MAC and network layers, reinforcement learning and deep learning models drive radio resource management, including power control, spectrum allocation, and interference coordination [72]. At the application layer, recurrent and transformer-based architectures forecast network traffic load [73], while graph neural networks detect intrusions in network telemetry.

Two architectural developments have accelerated this integration. The Open Radio Access Network (O-RAN) Alliance specification embeds a RAN Intelligent Controller (RIC) into the network architecture, providing standardized interfaces through which ML-based xApps and rApps execute closed-loop optimization [74, 75]. The AI-native network paradigm envisioned for 6G systems further elevates machine learning from a performance enhancement to a design primitive [76].

The consequence is that trained models in wireless systems are embedded control-plane and data-plane components whose predictions directly govern radio resource allocation and service-level agreement enforcement. A model whose parameters encode the influence of data that is private, malicious, or obsolete, therefore poses concrete operational risks: degraded throughput, violated latency guarantees, compromised user privacy, and regulatory non-compliance.

Unlearning Requirements in Wireless Deployments

The three categories of data risk formalized in section "Limitations of Machine Learning" each acquire distinctive characteristics when instantiated in wireless network environments, arising from the broadcast nature of the wireless medium, the mobility and heterogeneity of end devices, the stringent latency and reliability requirements of mission-critical services, and the massive scale of data generated by dense deployments.

Privacy Compliance at the Wireless Edge

Wireless networks generate vast quantities of user-attributable data (location trajectories, channel state information fingerprints, application usage patterns, device telemetry), all of which constitute personal data under the GDPR [8], the CCPA [9], and sector-specific telecommunications regulations such as the EU ePrivacy Directive [77]. Federated learning (FL) has emerged as the predominant privacy-aware training paradigm for wireless edge intelligence [78, 79]. However, FL does not eliminate data influence from the global model; gradient updates transmitted by each participating device are vulnerable to gradient inversion attacks [80] and membership inference [81].

When a mobile user withdraws consent or exercises a right to erasure, the influence of that user's data on the global federated model must be removed. Naïve retraining of the global model requires re-executing the entire federated training process across all remaining clients, a cost that is prohibitive for large-scale wireless deployments [82, 83]. These obligations impose the same verifiability requirements analyzed in section "Unlearning for Private and Copyrighted Data".

Security in Adversarial Wireless Environments

The broadcast nature of the wireless medium and the heterogeneity of connected devices create an expanded attack surface for data poisoning and backdoor attacks. In cognitive radio networks, adversaries controlling a subset of sensors can inject falsified spectrum reports [84]. In vehicular networks, compromised vehicles can inject poisoned sensor data to corrupt shared perception models [85]. In massive IoT deployments, resource-constrained devices are susceptible to firmware compromise, enabling backdoor-triggering data injection [86].

The security remediation scenario in wireless networks inherits the characteristics of section "Unlearning for Malicious and Corrupted Data" but adds two complications. First, malicious data points may be dispersed across multiple edge nodes, requiring coordinated unlearning across the distributed infrastructure. Second, safety-critical wireless services such as URLLC impose stringent time budgets on remediation [87]. These constraints favor the noise-injection-based unlearning of Chap. 6, which achieves low per-request cost while providing certified removal guarantees.

Adaptation to Non-Stationary Channels

Non-stationarity is the defining characteristic of wireless propagation environments. Wireless channels exhibit time-varying fading caused by user mobility, scatterer dynamics, and atmospheric conditions, with coherence times ranging from milliseconds to seconds [88]. This temporal variation induces concept drift (Definition 1.9) at rates that far exceed those encountered in most other domains. A channel estimator trained under one propagation regime encodes decision boundaries that become misaligned when the environment changes [69], and the historical training data actively degrades performance by anchoring parameters to an obsolete distribution.

Network-level reconfiguration events (dynamic cell splitting, beamforming codebook updates, network slice instantiation and teardown) introduce additional distributional shifts in O-RAN deployments [74]. These wireless-specific manifestations of concept drift correspond to the data maintenance scenario of section "Unlearning for Outdated, Biased, and Low-Quality Data", where the high frequency of distributional shifts demands unlearning algorithms with exceptionally low amortized per-request cost.

Design Constraints and Book Roadmap

Wireless network deployments impose a distinctive set of resource and architectural constraints that differentiate the unlearning design problem from its formulation in centralized settings. This subsection analyzes the three most consequential constraints (Section "Wireless-Specific Design Constraints") and maps the technical contributions of this book to the wireless application domains they address (Section "Chapter-To-Application Mapping").

Wireless-Specific Design Constraints

Three domain-specific constraints collectively shape the feasibility and design priorities of unlearning algorithms in wireless systems.

Edge Computation Limits.

Edge servers, small-cell base stations, and user equipment operate under stringent Size, Weight, and Power (SWaP) budgets that preclude the full retraining cost of $O(E \cdot N \cdot p)$ [89, 90]. In O-RAN deployments, near-real-time control loops impose tight time budgets (typically sub-second, e.g., on the order of 10 ms to 1 s for near-real-time RIC functions), making multi-hour retraining infeasible for operational remediation. The noise-injection-based unlearning of Chap. 6, which achieves $O(p)$ per-request cost with no gradient storage or Hessian computation, is designed to meet these constraints.

Distributed and Federated Architectures.

In wireless federated learning, the training data $\mathcal{D}$ is partitioned across K clients that never share raw data. A user-level unlearning request targeting client k must remove the influence of its local dataset from the global model without requiring remaining clients to retransmit data or re-execute local training [82]. The exact unlearning framework of Chap. 5, which achieves data removal through modular model components, provides a natural fit for this distributed setting. The data influence measurement frameworks of Chaps. 3 and 4 further support this architecture by quantifying each client's contribution to the global model.

Real-Time Service Guarantees.

A 5G URLLC service with a one-millisecond latency target and 10^{-5} reliability requirement cannot tolerate stale or compromised predictions while a multi-hour retraining job completes [91]. The checkpoint-based approaches of Chap. 4 and the partitioned-training approach of Chap. 5 both enable rapid rollback to a pre-contamination model state.

Chapter-To-Application Mapping

Table 1.2 summarizes the correspondence between the book's technical contributions and the wireless application domains they address. The theoretical foundations developed in Part I (Chaps. 2 through 4) provide the mathematical tools for quantifying and tracking data influence in the dynamic, distributed settings characteristic of wireless deployments. The algorithmic frameworks of Part II (Chaps. 5 and 6) translate these tools into practical unlearning methods whose computational profiles are compatible with the resource and latency constraints of modern wireless infrastructure.

Summary and Transition.

Machine unlearning arises with particular urgency in wireless network deployments: privacy compliance demands unlearning in federated edge systems, security remediation demands rapid decontamination under URLLC latency constraints, and adaptation to non-stationary channels demands continuous removal of obsolete data influence. The resource constraints of edge computing, the distributed architecture of federated wireless learning, and the real-time service guarantees of modern cellular systems collectively amplify the efficiency, scalability, and latency requirements for unlearning algorithms. Section "Structure of the Book" provides an overview of the book's organization.

Table 1.2 Mapping of book chapters to wireless network application domains and unlearning requirements

Chapter	Wireless application	Unlearning requirement
Chapter 2: Background and data influence analysis	Influence auditing for FL-based edge intelligence; fairness assessment in network resource allocation	Quantifying per-client or per-device contributions to the global model
Chapter 3: Online data influence measurement	Real-time detection of poisoned spectrum reports or corrupted IoT telemetry during training	Proactive identification of harmful data before influence becomes deeply embedded
Chapter 4: Time-varying influence measurement	Tracking temporal decay of CSI data relevance under channel drift; checkpoint-based rollback for RIC models	Retrospective identification of obsolete data for removal as the wireless environment evolves
Chapter 5: Exact unlearning via prompt tuning	User consent withdrawal in federated wireless networks; regulatory compliance for telecom operators	Verifiable, complete removal of a client's influence with reduced retraining cost
Chapter 6: Approximate unlearning via noise injection	Continuous adaptation of channel estimators and traffic predictors; rapid decontamination of backdoored beam management models	Low-latency, $O(p)$-cost certified removal compatible with edge SWaP constraints and URLLC deadlines

Structure of the Book

This book presents a unified framework for understanding and developing machine unlearning. It is organized into two main parts, each addressing complementary aspects of the field: the theory of unlearning and its practical applications.

Part I: Theory of Machine Unlearning.

Part I establishes the theoretical basis for machine unlearning by analyzing how data influence is encoded, measured, and removed in learning systems.

Chapter 1 introduces the motivation and scope of machine unlearning. It reviews the principles of traditional machine learning, highlights the irreversibility and limitations of data-driven optimization, and formalizes why the ability to selectively forget is both theoretically necessary and practically important.

Chapter 2 provides the background and formal problem definition for machine unlearning. It examines classical data influence analysis methods and their limitations, surveys exact and approximate unlearning techniques, and establishes the evaluation framework used throughout the book.

Chapter 3 develops the theory of *online data influence measurement*, formulating a real-time framework for quantifying the effect of each data point on model optimization. This theory supports proactive unlearning by identifying harmful or low-utility data during training, before their influence becomes deeply embedded.

Chapter 4 extends the framework to *time-varying data influence measurement*, which captures cumulative and temporally localized effects of data points across training intervals. This formulation introduces the propagator matrix $Z_t = I - \eta_t H^{[t]}$ whose spectral properties govern how influence evolves through subsequent optimization steps, a structure that directly underpins the noise-injection-based unlearning of Chap. 6.

Together, these chapters form a coherent theoretical foundation connecting optimization dynamics, influence estimation, and information removal into a unified mathematical theory of unlearning.

Part II: Applications of Machine Unlearning.

Part II translates the theoretical insights of Part I into algorithmic frameworks suitable for modern learning systems.

Chapter 5 introduces LMEraser, an exact unlearning method for large-scale vision models. LMEraser partitions training data into public and private subsets, freezes the backbone after pre-training on public data, and adaptively clusters private data to train dedicated prompts and classifier heads for each cluster. Unlearning requires re-tuning only the affected cluster's prompt and classifier head, achieving complete data removal with a 100-fold reduction in unlearning cost relative to single-prompt methods.

Chapter 6 presents a certified unlearning framework that operates during training via decay-aware noise injection. Building on the propagator analysis of Chap. 4, it shows that weight decay causes each data point's influence to decay exponentially through subsequent optimization steps, enabling (ε, δ)-certified unlearning through calibrated Gaussian noise injection at $O(p)$ per-request cost, with no gradient storage, Hessian inversion, or trajectory replay.

Chapter 7 concludes the book with a discussion of open challenges, evaluation frameworks, and future directions, with particular attention to the distinctive requirements of wireless network deployments.

Summary.

Overall, this book progresses from theory to practice: data influence measurement (Chaps. 3 and 4) provides the mathematical tools for understanding how training data shapes model behavior, while exact unlearning via prompt tuning (Chap. 5) and certified unlearning via noise injection (Chap. 6) translate these tools into practical removal mechanisms. By combining theoretical development with practical algorithms, the book positions machine unlearning not as an afterthought but as a natural extension of machine learning itself.

References

1. Goodfellow, I., Bengio, Y., Courville, A.: Deep Learning. MIT Press. http://www.deeplearningbook.org (2016)
2. Bishop, C.M., Nasrabadi, N.M.: Pattern Recognition and Machine Learning, Vol. 4. Springer (2006)
3. Hastie, T., Tibshirani, R., Friedman, J.: The elements of statistical learning (2009)
4. Moreno-Torres, J.G. et al.: A unifying view on dataset shift in classification. Pattern Recogn. **45**(1), 521–530 (2012)
5. Natarajan, N. et al.: Learning with noisy labels. Adv. Neural Inform. Process. Syst. **26** (2013)
6. He, H., Garcia, E.A.: Learning from imbalanced data. IEEE Trans. Knowl. Data Eng. **21**(9), 1263–1284 (2009)
7. Hendrycks, D., Dietterich, T.: Benchmarking neural network robustness to common corruptions and perturbations. arXiv preprint arXiv:1903.12261 (2019)
8. Voigt, P., Von dem Bussche, A.: The EU General Data Protection Regulation (GDPR): A Practical Guide (2017). https://doi.org/10.1007/978-3-319-57959-7
9. California Consumer Privacy Act of 2018: California Legislative Information, AB-375 (2018)
10. Biggio, B., Nelson, B., Laskov, P., et al.: Poisoning attacks against support vector machines. In: Proceedings of the 29th International Conference on Machine Learning, ICML 2012, pp. 1807–1814 (2012)
11. Shafahi, A., et al.: Poison frogs! targeted clean-label poisoning attacks on neural networks. Adv. Neural Inform. Process. Syst. **31** (2018)
12. Shalev-Shwartz, S., Ben-David, S.: Understanding Machine Learning: From Theory to Algorithms. Cambridge University Press (2014)
13. Mohri, M., Rostamizadeh, A., Talwalkar, A.: Foundations of Machine Learning. MIT press (2018)
14. He, K., et al.: Deep residual learning for image recognition. In: Proceedings of the IEEE Conference on Computer Vision and Pattern Recognition, pp. 770–778 (2016)
15. Vaswani, A., et al.: Attention is all you need. Adv. Neural Inform. Process. Syst. **30** (2017)
16. Hinton, G., et al.: Deep neural networks for acoustic modeling in speech recognition: the shared views of four research groups. IEEE Signal Process. Mag. **29**(6), 82–97 (2012)
17. Zhang, C., et al.: Understanding deep learning (still) requires rethinking generalization. Commun. ACM **64**(3), 107–115 (2021)
18. Belkin, M., et al.: Reconciling modern machine-learning practice and the classical bias–variance trade-off. Proc. Natl. Acad. Sci. **116**(32), 15849–15854 (2019)
19. Carlini, N., et al.: Extracting training data from large language models. In: 30th USENIX Security Symposium (USENIX Security 21), pp. 2633–2650 (2021)
20. Krogh, A., Hertz, J.: A simple weight decay can improve generalization. Adv. Neural Inform. Process. Syst. **4** (1991)
21. Tibshirani, R.: Regression shrinkage and selection via the lasso. J. Roy. Stat. Soc. Ser. B: Stat. Methodol. **58**(1), 267–288 (1996)
22. Bottou, L., Curtis, F.E., Nocedal, J.: Optimization methods for large-scale machine learning. SIAM Rev. **60**(2), 223–311 (2018)
23. Robbins, H., Monro, S.: A stochastic approximation method. Ann. Math. Stat. 400–407 (1951)
24. Polyak, B.T.: Some methods of speeding up the convergence of iteration methods. Ussr Comput. Math. Math. Phys. **4**(5), 1–17 (1964)
25. Duchi, J., Hazan, E., Singer, Y.: Adaptive subgradient methods for online learning and stochastic optimization. J. Mach. Learn. Res. **12**(7) (2011)
26. Kingma, D.P., Ba, J.: Adam: a method for stochastic optimization. arXiv preprint arXiv:1412.6980 (2014)
27. Pruthi, G., et al.: Estimating training data influence by tracing gradient descent. Adv. Neural. Inf. Process. Syst. **33**, 19920–19930 (2020)
28. Bousquet, O., Elisseeff, A.: Stability and generalization. J. Mach. Learn. Res. **2**, 499–526 (2002)

29. Shokri, R., et al.: Membership inference attacks against machine learning models. In: IEEE Symposium on Security and Privacy (SP), pp. 3–18. IEEE (2017)
30. Thudi, A., et al.: Unrolling SGD: understanding factors influencing machine unlearning. In: IEEE 7th European Symposium on Security and Privacy (EuroS&P), pp. 303–319. IEEE (2022)
31. Nguyen, T.T., et al.: A survey of machine unlearning. arXiv preprint arXiv:2209.02299 (2022)
32. Xu, J., et al.: Machine unlearning: solutions and challenges. IEEE Trans. Emerg. Top. Comput. Intell. **8**(3), 2150–2168 (2024)
33. Kohavi, R., et al.: A study of cross-validation and bootstrap for accuracy estimation and model selection. In: Ijcai, Vol. 14. 2, Montreal, Canada, pp. 1137–1145 (1995)
34. Paleyes, A., Urma, R.-G., Lawrence, N.D.: Challenges in deploying machine learning: a survey of case studies. ACM Comput. Surv. **55**(6), 1–29 (2022)
35. Sculley, D., et al.: Hidden technical debt in machine learning systems. Adv. Neural Inform. Process. Syst. **28** (2015)
36. Brown, T., et al.: Language models are few-shot learners. Adv. Neural. Inf. Process. Syst. **33**, 1877–1901 (2020)
37. Chowdhery, A., et al.: Palm: scaling language modeling with pathways. J. Mach. Learn. Res. **24**(240), 1–113 (2023)
38. Lu, J., et al.: Learning under concept drift: a review. IEEE Trans. Knowl. Data Eng. **31**(12), 2346–2363 (2018)
39. Gama, J., et al.: A survey on concept drift adaptation. ACM Comput. Surv. (CSUR) **46**(4), 1–37 (2014)
40. Parisi, G.I., et al.: Continual lifelong learning with neural networks: a review. Neural Netw. **113**, 54–71 (2019)
41. Fredrikson, M., Jha, S., Ristenpart, T.: Model inversion attacks that exploit confidence information and basic countermeasures. In: Proceedings of the 22nd ACM SIGSAC Conference on Computer and Communications Security, pp. 1322–1333 (2015)
42. Carlini, N., et al.: Membership inference attacks from first principles. In: IEEE Symposium on Security and Privacy (SP), pp. 1897–1914. IEEE (2022)
43. Zhang, Y., et al.: The secret revealer: generative model-inversion attacks against deep neural networks. In: Proceedings of the IEEE/CVF Conference on Computer Vision and Pattern Recognition, pp. 253–261 (2020)
44. Carlini, N., et al.: Quantifying memorization across neural language models. In: The Eleventh International Conference on Learning Representations (2022)
45. Feldman, V.: Does learning require memorization? a short tale about a long tail. In: Proceedings of the 52nd Annual ACM SIGACT Symposium on Theory of Computing, pp. 954–959 (2020)
46. Ginart, A., et al.: Making AI forget you: Data deletion in machine learning. Adv. Neural Inform. Process. Syst. **32** (2019)
47. Vale de Castro, E.T., Silva, G.R.S., Canedo, E.D.: Ensuring privacy in the application of the Brazilian general data protection law (LGPD). In: Proceedings of the 37th ACM/SIGAPP Symposium on Applied Computing, pp. 1228–1235 (2022)
48. Calzada, I.: Citizens' data privacy in China: The state of the art of the Personal Information Protection Law (PIPL). Smart Cities **5**(3), 1129–1150 (2022)
49. Ministry of Electronics and Information Technology, Government of India: The Digital Personal Data Protection Act, 2023. Act No. 22 of 2023, Gazette of India, 11 August 2023. Aug. 2023. https://www.meity.gov.in/static/uploads/2024/06/2bf1f0e9f04e6fb4f8fef35e82c42aa5.pdf (visited on 03 May 2026)
50. Steinhardt, J., Koh, P.W.W., Liang, P.S.: Certified defenses for data poisoning attacks. Adv. Neural Inform. Process. Syst. **30** (2017)
51. Gu, T., Dolan-Gavitt, B., Garg, S.: Badnets: identifying vulnerabilities in the machine learning model supply chain. arXiv preprint arXiv:1708.06733 (2017)
52. Liu, Y., et al.: Trojaning attack on neural networks. In: 25th Annual Network and Distributed System Security Symposium (NDSS 2018). Internet Soc (2018)
53. Li, Y., et al.: Invisible backdoor attack with sample-specific triggers. In: Proceedings of the IEEE/CVF International Conference on Computer Vision, pp. 16463–16472 (2021)

54. Nguyen, A., Tran, A.: Wanet–imperceptible warping-based backdoor attack. arXiv preprint arXiv:2102.10369 (2021)
55. Northcutt, C.G., Athalye, A., Mueller, J.: Pervasive label errors in test sets destabilize machine learning benchmarks. arXiv preprint arXiv:2103.14749 (2021)
56. Koh, P.W., Liang, P.: Understanding black-box predictions via influence functions. In: International Conference on Machine Learning, pp. 1885–1894. PMLR (2017)
57. Hardt, M., Price, E., Srebro, N.: Equality of opportunity in supervised learning. Adv. Neural Inform. Process. Syst. **29** (2016)
58. Barocas, S., Selbst, A.D.: Big data's disparate impact. Calif. L. Rev. **104**, 671 (2016)
59. Mehrabi, N., et al.: A survey on bias and fairness in machine learning. ACM Comput. Surv. (CSUR) **54**(6), 1–35 (2021)
60. Obermeyer, Z., et al.: Dissecting racial bias in an algorithm used to manage the health of populations. Science **366**(6464), 447–453 (2019)
61. Warnecke, A., et al.: Machine unlearning of features and labels. In: "30th Annual Network and Distributed System Security Symposium, –NDSS" 2023, San Diego, California, USA, February 27–March 3, 2023. The Internet Society (2023). https://www.ndss-symposium.org/ndss-paper/machine-unlearning-of-features-and-labels/
62. Bourtoule, L., et al.: Machine unlearning. In: IEEE Symposium on Security and Privacy (SP), pp. 141–159. IEEE (2021)
63. Guo, C., et al.: Certified data removal from machine learning models. In: International Conference on Machine Learning, pp. 3832–3842. PMLR (2020)
64. Golatkar, A., Achille, A., Soatto, S.: Eternal sunshine of the spotless net: selective forgetting in deep networks. In: Proceedings of the IEEE/CVF Conference on Computer Vision and Pattern Recognition, pp. 9304–9312 (2020)
65. Neel, S., Roth, A., Sharifi-Malvajerdi, S.: Descent-to-delete: gradient-based methods for machine unlearning. In: Algorithmic Learning Theory, pp. 931–962. PMLR (2021)
66. Dwork, C., et al.: Calibrating noise to sensitivity in private data analysis. In: Theory of Cryptography Conference, pp. 265–284. Springer, Berlin (2006)
67. Eldan, R., Russinovich, M.: Who's Harry Potter? approximate unlearning in LLMs (2023). arXiv:2310.02238
68. Jang, J., et al.: Knowledge unlearning for mitigating privacy risks in language models. In: Proceedings of the 61st Annual Meeting of the Association for Computational Linguistics (Volume 1: Long Papers), pp. 14389–14408 (2023)
69. Soltani, M., et al.: Deep learning-based channel estimation. IEEE Commun. Lett. **23**(4), 652–655 (2019)
70. Hassan, H.A., et al.: Effective deep learning-based channel state estimation and signal detection for OFDM wireless systems. J. Electr. Eng. **74**(3), 167–176 (2023)
71. Simeone, O.: A very brief introduction to machine learning with applications to communication systems. IEEE Trans. Cogn. Commun. Netw. **4**(4), 648–664 (2018)
72. Mao, Q., Hu, F., Hao, Q.: Deep learning for intelligent wireless networks: a comprehensive survey. IEEE Commun. Surv. Tutor. **20**(4), 2595–2621 (2018)
73. Wang, J., et al.: Thirty years of machine learning: the road to Pareto-optimal wireless networks. IEEE Commun. Surv. Tutor. **22**(3), 1472–1514 (2020)
74. Polese, M., et al.: Understanding O-RAN: architecture, interfaces, algorithms, security, and research challenges. IEEE Commun. Surv. Tutor. **25**(2), 1376–1411 (2023)
75. Bonatim, L., et al.: Intelligence and learning in O-RAN for data-driven NextG cellular networks. IEEE Commun. Mag. **59**(10), 21–27 (2021)
76. Letaief, K.B., et al.: The roadmap to 6G: AI empowered wireless networks. IEEE Commun. Mag. **57**(8), 84–90 (2019)
77. Bond, R.: The EU e-Privacy directive and consent to cookies. Bus. Lawyer 215–223 (2012)
78. McMahan, B., et al.: Communication-efficient learning of deep networks from decentralized data. In: Artificial Intelligence and Statistics, pp. 1273–1282. PMLR (2017)
79. Lim, W.Y.B., et al.: Federated learning in mobile edge networks: a comprehensive survey. IEEE Commun. Surv. Tutor. **22**(3), 2031–2063 (2020)

80. Zhu, L., Liu, Z., Han, S.: Deep leakage from gradients. In: Advances in Neural Information Processing Systems, vol. 32 (2019)
81. Nasr, M., Shokri, R., Houmansadr, A.: Comprehensive privacy analysis of deep learning: passive and active white-box inference attacks against centralized and federated learning. In: IEEE Symposium on Security and Privacy (SP), pp. 739–753. IEEE (2019)
82. Wu, C., et al.: A federated graph neural network framework for privacy-preserving personalization. Nat. Commun. **13**(1), 3091 (2022)
83. Liu, G., et al.: Federaser: enabling efficient client-level data removal from federated learning models. In: IEEE/ACM 29th International Symposium on Quality of Service (IWQOS), pp. 1–10. IEEE (2021)
84. Sagduyu, Y.E., Erpek, T., Shi, Y.: Adversarial machine learning for 5G communications security. In: Game Theory and Machine Learning for Cyber Security, pp. 270–288 (2021)
85. Tu, J., et al.: Adversarial attacks on multi-agent communication. In: Proceedings of the IEEE/CVF International Conference on Computer Vision, pp. 7768–7777 (2021)
86. Nguyen, T.N., et al.: Poisoning attacks on federated learning-based IoT intrusion detection system. In: Proceedings of the Workshop Decentralized IoT System Security (DISS), vol. 79, pp. 1–7 (2020)
87. Siddiqui, M.U.A., et al.: URLLC in beyond 5G and 6G networks: an interference management perspective. IEEe Access **11**, 54639–54663 (2023)
88. Tse, D., Viswanath, P.: Fundamentals of Wireless Communication. Cambridge University Press (2005)
89. Chen, J., Ran, X.: Deep learning with edge computing: a review. Proc. IEEE **107**(8), 1655–1674 (2019)
90. Li, E., et al.: Edge AI: on-demand accelerating deep neural network inference via edge computing. IEEE Trans. Wirel. Commun. **19**(1), 447–457 (2019)
91. Bennis, M., Debbah, M., Poor, H.V.: Ultrareliable and low-latency wireless communication: tail, risk, and scale. Proc. IEEE **106**(10), 1834–1853 (2018)

Chapter 2
Background of Machine Unlearning

Abstract This chapter surveys the technical foundations underlying machine unlearning: data influence analysis and existing unlearning methods. For influence analysis, we examine four method families (influence functions, game-theoretic valuation, training trajectory methods, and surrogate approaches) and identify five persistent limitations concerning scalability, violated assumptions, temporal dynamics, inter-data interactions, and approximation accuracy. For unlearning methods, we organize the landscape into exact approaches that achieve distributional equivalence through data partitioning and isolated retraining, and approximate approaches that trade strict equivalence for efficiency through parameter manipulation, gradient ascent, knowledge editing, or output filtering. We detail the assumptions each method requires and the settings where those assumptions break down. The chapter concludes with the evaluation framework for machine unlearning, covering forgetting quality metrics, utility preservation metrics, and standardized benchmarks including TOFU, MUSE, WMDP, and OpenUnlearning.

Data Influence Analysis

Classical Data Influence Analysis

Data influence analysis encompasses a family of techniques that quantify how individual training data points affect a model's learned parameters, predictions, and generalization behavior. Chapter 1 established that every data point $z_i \in \mathcal{D}_{\text{train}}$ contributes to the empirical risk (Eq. (1.1)) and shapes the optimization trajectory (Eq. (1.7)), encoding its influence in the final parameters $\boldsymbol{\theta}^*$ in a manner that is generally nonlinear and non-decomposable. Data influence analysis seeks to disentangle this aggregate influence by assigning to each data point a scalar score that reflects its contribution to a specified model property, such as the loss on a particular test data point, the model's overall accuracy, or a fairness metric. The resulting scores serve diverse purposes: identifying mislabeled or corrupted data points, selecting informative subsets for efficient training, attributing model behavior to specific training inputs, and guiding the selective removal of data influence in machine unlearning.

J. Xu and X. Jia, *Machine Unlearning*, Wireless Networks,
https://doi.org/10.1007/978-3-032-30978-5_2

Before surveying specific methods, it is useful to distinguish two fundamentally different approaches to measuring data influence. The first approach, *LOO-based methods*, treats the Leave-One-Out (LOO) loss change as the gold-standard measure of data influence: how much does the model's loss on a test data point (or on the full test set) change when a specific training data point is removed and the model is retrained? LOO retraining computes this quantity exactly by performing the removal and retraining, and produces scores that are interpretable as actual loss differences with concrete units tied to the loss function. Influence Functions (IFs) analytically approximate the same LOO loss change through local curvature information at the converged model. Both methods target the same physical quantity, differing only in whether it is measured exactly or approximated.

However, LOO retraining requires N full retraining runs and is therefore computationally prohibitive for large-scale models. IFs avoid retraining but impose strong assumptions, including loss convexity and training convergence, that are rarely satisfied in modern deep learning, and their computational cost remains substantial for high-dimensional parameter spaces. These limitations have motivated a second approach: *custom influence scoring* methods that bypass LOO approximation entirely and instead define efficient, task-oriented scores tailored to specific application needs. Rather than estimating the actual loss change from data removal, each family of custom scoring methods designs its influence metric around a distinct practical objective.

Game-theoretic valuation methods are motivated by the need for *fair credit allocation* in settings such as data marketplaces and collaborative learning, where multiple data contributors supply training data and require a principled mechanism for distributing compensation or attributing value [1, 2]. The Shapley value's axiomatic fairness guarantees (efficiency, symmetry, null player, linearity) ensure that each contributor's payment reflects their genuine contribution, making game-theoretic scores the natural choice when equitable data valuation is the primary concern. Training trajectory methods are motivated by *data quality assessment* tasks such as data cleansing, corrupted-data detection, and data pruning [3, 4]. The key insight is that a data point's gradient contribution during training, including its magnitude, its alignment with the overall optimization direction, and its evolution across epochs, directly reflects how the data point drives parameter updates. Abnormally large or misaligned gradients signal corrupted or mislabeled data points, while consistently small gradients indicate redundant data points that can be pruned without performance loss, making gradient-derived scores natural proxies for data quality. Surrogate and representation-based methods are motivated by *scalable training data attribution* for model debugging and auditing [5, 6]. When a deployed model produces an unexpected or erroneous prediction, practitioners need to trace the prediction back to the responsible training data points. By learning an explicit mapping from data composition to model output through auxiliary models, surrogate methods can efficiently answer "which training data points are responsible for this prediction?" across many queries without retraining, making them particularly suited to post-hoc model interpretability and systematic auditing workflows.

This subsection surveys both approaches across four method families. LOO retraining and IFs (section "Leave-One-Out Retraining and Influence Functions") constitute the LOO approximation approach, establishing the gold standard and its principal analytical approximation. The remaining three families belong to the custom influence scoring approach, each designed around the distinct application objectives described above: game-theoretic valuation methods (section "Game-Theoretic Valuation Methods") for fair credit allocation, training trajectory methods (section "Training Trajectory Methods") for data quality assessment, and surrogate and representation-based methods (section "Surrogate and Representation-Based Methods") for scalable training data attribution. For each family, we formalize the core methodology, review key developments and extensions, and identify the assumptions that govern applicability. The systematic limitations of these methods, which motivate the influence measurement frameworks developed in Chaps. 3 and 4, are analyzed in section "Limitations of Classical Methods".

Leave-One-Out Retraining and Influence Functions

Leave-One-Out (LOO) retraining and Influence Functions (IFs) both target the same quantity: the change in model loss that results from removing a single training data point and retraining. LOO retraining computes this quantity exactly by performing the removal and retraining, while IFs approximate it analytically using the local curvature of the loss landscape at the converged model. We treat them together because IFs are, by construction, first-order approximations to the LOO loss change, and their accuracy is evaluated by comparison to LOO ground truth.

Leave-One-Out Retraining.

LOO retraining is the oldest and most direct method for measuring data influence, originating in the classical regression diagnostics of Cook [7] and subsequently generalized to arbitrary learning algorithms [8, 9]. The method isolates the effect of a single data point by comparing a model trained on the full dataset to a model trained on the dataset with that data point removed.

Definition 2.1 (*Leave-One-Out Influence*) Let $\mathcal{A}$ be a training algorithm, $\boldsymbol{\theta}^* = \mathcal{A}(\mathcal{D})$ the parameters trained on the full dataset $\mathcal{D}$, and $\boldsymbol{\theta}^*_{-i} = \mathcal{A}(\mathcal{D} \setminus \{z_i\})$ the parameters trained on $\mathcal{D}$ with data point z_i removed. The *LOO influence* of z_i on a test data point $z_{\text{test}} = (\mathbf{x}_{\text{test}}, y_{\text{test}})$ is defined as:

$$\mathcal{I}_{\text{LOO}}(z_i, z_{\text{test}}) = \ell\big(h_{\boldsymbol{\theta}^*}(\mathbf{x}_{\text{test}}), y_{\text{test}}\big) - \ell\big(h_{\boldsymbol{\theta}^*_{-i}}(\mathbf{x}_{\text{test}}), y_{\text{test}}\big). \tag{2.1}$$

A positive value indicates that including z_i in training *increased* the loss on z_{test} (a harmful influence), while a negative value indicates that z_i *decreased* the test loss (a beneficial influence).

The LOO influence can also be defined with respect to aggregate performance metrics rather than individual test data points. Replacing the test-point-specific loss with the average test loss yields the *global LOO influence*:

$$\mathcal{I}_{\text{LOO}}^{\text{global}}(z_i) = \hat{R}(\boldsymbol{\theta}^*;\ \mathcal{D}_{\text{test}}) - \hat{R}(\boldsymbol{\theta}^*_{-i};\ \mathcal{D}_{\text{test}}), \tag{2.2}$$

which measures the overall effect of z_i on model generalization. A closely related quantity is the *self-influence* $\mathcal{I}_{\text{LOO}}(z_i, z_i)$, which measures how the inclusion of z_i in training affects the model's prediction on z_i itself and is closely related to the memorization behavior analyzed in section "Key Components of Machine Learning".

LOO retraining serves as the gold standard for data influence measurement because it makes no approximations: the influence is computed by directly observing the outcome of removing the data point and retraining the model. The resulting score has a concrete physical interpretation: it is the actual loss change, measured in the same units as the loss function ℓ, that would occur if the data point were removed and the model fully retrained. However, this exactness comes at a prohibitive computational cost. Evaluating the LOO influence for all N training data points requires N full retraining runs, each costing $O(E \cdot N \cdot p)$ (section "The Process of Machine Learning"), yielding a total complexity of $O(N^2 \cdot E \cdot p)$. For modern deep networks where N may reach 10^6 and a single training run may require days or weeks of computation, this cost is entirely impractical. Furthermore, LOO retraining evaluates data points in isolation and therefore cannot capture the combinatorial interactions between data points, such as the synergistic or antagonistic effects that arise when multiple data points are jointly added or removed [10]. A further complication revealed by recent empirical studies [10–12] is that LOO influence estimates are sensitive to the randomness inherent in Stochastic Gradient Descent (SGD) training: independent LOO retraining runs on the same data can produce different models and hence different influence scores, making the LOO influence itself a random variable rather than a deterministic quantity. These limitations motivate IFs as an analytical approximation.

Influence Function.

IFs provide an analytical approximation to the LOO influence by leveraging the local curvature of the loss landscape at the converged model, avoiding the need for retraining [9]. Originally developed in the robust statistics literature for diagnosing influential observations in linear models [7], IFs were adapted to the machine learning setting by Koh and Liang [9], who demonstrated their applicability to logistic regression and, with caveats, to deep neural networks.

The derivation proceeds by considering the effect of infinitesimally upweighting a single training data point z_i in the empirical risk. Define the perturbed parameters as the minimizer of the upweighted objective:

$$\boldsymbol{\theta}^*_{\epsilon, z_i} = \arg\min_{\boldsymbol{\theta} \in \Theta}\ \hat{R}(\boldsymbol{\theta}; \mathcal{D}) + \epsilon\, \ell\big(h_{\boldsymbol{\theta}}(\mathbf{x}_i), y_i\big), \tag{2.3}$$

where ϵ controls the perturbation magnitude. Setting $\epsilon = -1/N$ corresponds (to first order) to removing z_i from the dataset. Applying the implicit function theorem to the first-order optimality condition $\nabla_{\boldsymbol{\theta}}[\hat{R}(\boldsymbol{\theta}; \mathcal{D}) + \epsilon\, \ell(h_{\boldsymbol{\theta}}(\mathbf{x}_i), y_i)] = \mathbf{0}$ yields the influence of data point z_i on the optimal parameters:

$$\left.\frac{d\boldsymbol{\theta}^*_{\epsilon,z_i}}{d\epsilon}\right|_{\epsilon=0} = -\mathbf{H}_{\theta^*}^{-1} \nabla_\theta \ell\big(h_{\theta^*}(\mathbf{x}_i), y_i\big), \tag{2.4}$$

where $\mathbf{H}_{\theta^*}$ is the Hessian of the empirical risk evaluated at the converged parameters:

$$\mathbf{H}_{\theta^*} = \frac{1}{N} \sum_{j=1}^{N} \nabla_\theta^2 \ell\big(h_{\theta^*}(\mathbf{x}_j), y_j\big). \tag{2.5}$$

The influence on a test data point z_{test} is then obtained by applying the chain rule to propagate the parameter perturbation through the test loss, yielding the classical IF.

Definition 2.2 (*Influence Function*) The *Influence Function* (IF) estimate of the effect of training data point z_i on the loss at test data point z_{test} is:

$$\mathcal{I}_{\text{IF}}(z_i, z_{\text{test}}) = -\nabla_\theta \ell\big(h_{\theta^*}(\mathbf{x}_{\text{test}}), y_{\text{test}}\big)^\top \mathbf{H}_{\theta^*}^{-1} \nabla_\theta \ell\big(h_{\theta^*}(\mathbf{x}_i), y_i\big). \tag{2.6}$$

This quantity approximates the LOO influence: $\mathcal{I}_{\text{IF}}(z_i, z_{\text{test}}) \approx -\frac{1}{N} \mathcal{I}_{\text{LOO}}(z_i, z_{\text{test}})$.

The critical property that connects IFs to LOO is that both target the same quantity: the change in test loss caused by removing a specific data point. The IF achieves this by linearizing the dependence of the optimal parameters on the data composition, replacing the actual retraining with a first-order Taylor expansion around the converged model. This linearization is exact when the empirical risk is quadratic in $\boldsymbol{\theta}$ (as in linear regression with squared loss) and provides a good approximation when the loss landscape is locally well-approximated by a quadratic, which requires three critical assumptions [9, 13]. First, the loss function must be twice differentiable with respect to $\boldsymbol{\theta}$. Second, the empirical risk must be strictly convex in $\boldsymbol{\theta}$, ensuring that the Hessian $\mathbf{H}_{\theta^*}$ is positive definite and hence invertible. Third, the training algorithm must have converged to the global minimizer $\boldsymbol{\theta}^*$, so that the first-order optimality condition holds exactly. When these assumptions are satisfied (as in regularized logistic regression or ridge regression), the IF provides an accurate and computationally efficient approximation to LOO: computing $\mathcal{I}_{\text{IF}}(z_i, z_{\text{test}})$ requires a single inverse-Hessian-vector product rather than a full retraining run.

Scaling Influence Functions to Deep Neural Networks.

The direct computation of the IF in Eq. (2.6) is impractical for deep neural networks, where the parameter count p may exceed 10^9. Forming and inverting the $p \times p$ Hessian matrix requires $O(p^2)$ storage and $O(p^3)$ computation, costs that are infeasible at scale. A series of techniques has been developed to approximate the inverse-Hessian-Vector Product (HVP) $\mathbf{H}_{\theta^*}^{-1}\mathbf{v}$ without explicitly forming $\mathbf{H}_{\theta^*}$.

Koh and Liang [9] proposed using stochastic estimation via the **Li**near time **S**tochastic **S**econd-order **A**lgorithm (LiSSA) [14], which iteratively approximates the inverse-HVP through a truncated Neumann series using stochastic HVPs. Each HVP can be computed efficiently via automatic differentiation at a cost of $O(p)$ per iteration [15], avoiding the need to materialize the full Hessian. FastIF [16] introduced query batching and caching strategies to accelerate influence computation when evaluating the influence of many training data points on a fixed set of test data points, reducing wall-clock time by an order of magnitude over the original implementation. Schioppa et al. [17] proposed an Arnoldi iteration-based approach that projects the Hessian onto a low-dimensional Krylov subspace, enabling scalable influence estimation for models with hundreds of millions of parameters. Most recently, Grosse et al. [18] applied the Eigenvalue-corrected Kronecker-Factored Approximate Curvature (EK-FAC) method to approximate the Hessian for large language models, demonstrating IF computation at unprecedented scale.

Empirical Reliability and Failure Modes.

Despite these computational advances, the theoretical assumptions underlying IFs are rarely satisfied in modern deep learning, and a growing body of empirical evidence has documented significant discrepancies between IF estimates and true LOO influence [13, 19]. Basu et al. [13] demonstrated that IF estimates diverge from true LOO influence in non-convex settings, with the approximation quality degrading as the model departs from strict convexity. The fundamental issue is that deep neural networks have non-convex loss landscapes with multiple local minima, and the first-order Taylor approximation in Eq. (2.4) captures only the local curvature at a single point, which may not reflect the global effect of removing a data point and retraining.

Bae et al. [20] provided a more detailed diagnosis, identifying two specific failure modes. First, *proximal mismatch*: the IF estimates the parameter change due to an infinitesimal perturbation, but LOO retraining involves a finite (discrete) removal, and the discrepancy grows with the magnitude of the removal effect. Second, *warm-start bias*: practical LOO experiments often initialize the retrained model from the original parameters $\boldsymbol{\theta}^*$ (warm-start) rather than from a fresh random initialization (cold-start), and the resulting model $\boldsymbol{\theta}^*_{-i}$ may converge to a different local minimum than the cold-start retrained model, introducing a systematic discrepancy that the IF cannot capture. Basu et al. [21] proposed second-order corrections to mitigate the proximal mismatch, computing higher-order terms in the Taylor expansion to improve accuracy for larger perturbations. Schioppa et al. [22] provided a theoretical analysis establishing conditions under which the approximation error can be bounded. Choe et al. [23] investigated the impact of damping terms, commonly added to regularize the Hessian inversion, showing that the choice of damping parameter significantly affects the quality of influence estimates and proposing principled selection strategies.

The computational and theoretical limitations of LOO retraining and IFs, prohibitive retraining cost for the former and restrictive assumptions for the latter, have motivated the development of custom influence scoring methods that prioritize computational efficiency and practical utility over strict LOO approximation. The remaining three method families in this subsection belong to this second category.

Game-Theoretic Valuation Methods

LOO-based methods suffer from several fundamental limitations beyond computational cost. LOO retraining evaluates each data point in isolation: the influence score $\mathcal{I}_{\text{LOO}}(z_i, z_{\text{test}})$ measures the effect of removing z_i from the complete dataset $\mathcal{D}$, but it cannot capture the *combinatorial interactions* between data points [10]. In practice, data points often exhibit synergistic effects (two individually low-value data points that jointly provide essential coverage of a decision boundary) or antagonistic effects (two individually high-value data points that are redundant with each other), and LOO scores are blind to these interactions because they consider only the singleton removal $\mathcal{D} \setminus \{z_i\}$. Furthermore, LOO scores lack *fairness guarantees*: two data points with genuinely symmetric contributions to the model may receive different LOO scores due to the stochastic randomness of training [11], and the N individual LOO scores bear no principled relationship to the total model utility, as they do not sum to any meaningful aggregate quantity. These limitations are particularly problematic in settings such as data marketplaces, collaborative learning, and federated data contribution, where multiple data providers supply training data and require a principled, equitable mechanism for distributing compensation or attributing credit.

Game-theoretic valuation methods address these limitations by modeling the training process as a cooperative game in which data points are players and model performance is the payoff [1, 2]. Rather than evaluating each data point against a single reference subset, these methods assess each data point's marginal contribution across *all* possible subsets of the training data, thereby capturing inter-data interactions that LOO misses. The resulting scores satisfy rigorous axiomatic properties inherited from cooperative game theory, providing formal fairness guarantees that are absent from LOO-based approaches. The foundational framework is the Shapley value, and subsequent work has introduced alternative game-theoretic solution concepts, notably the Banzhaf value and its weighted variants, that relax certain axioms to achieve better computational or statistical properties.

Shapley Value Framework.

The Shapley value, originating in cooperative game theory [24], provides the foundational framework for game-theoretic data valuation. When applied to data valuation, each training data point is treated as a player, and a utility function measures model performance on a held-out validation set [1, 2].

Definition 2.3 (*Shapley Value*) Let $v : 2^{\mathcal{D}} \to \mathbb{R}$ be a utility function that maps a subset of the training data to a scalar performance measure (e.g., validation accuracy). The *Shapley value* of data point $z_i \in \mathcal{D}$ is:

$$\phi_i(v) = \frac{1}{N} \sum_{k=0}^{N-1} \binom{N-1}{k}^{-1} \sum_{\substack{S \subseteq \mathcal{D} \setminus \{z_i\} \\ |S|=k}} \big[v(S \cup \{z_i\}) - v(S)\big], \tag{2.7}$$

where the inner sum averages over all subsets of size k that exclude z_i, and the outer sum averages over all possible subset sizes. The quantity $v(S \cup \{z_i\}) - v(S)$ is the *marginal contribution* of z_i to subset S.

The Shapley value is the unique attribution method satisfying four axioms that are desirable for fair data valuation [1, 24]. *Efficiency* requires that the sum of all data points' Shapley values equals the difference between the full-data utility and the empty-set utility, $\sum_{i=1}^{N} \phi_i = v(\mathcal{D}) - v(\emptyset)$, ensuring that the total model utility is fully allocated among contributors. *Symmetry* requires that two data points with identical marginal contributions to every subset receive equal values. *Linearity* requires that the Shapley value of a linear combination of utility functions equals the corresponding linear combination of individual Shapley values. *Null player* requires that a data point whose marginal contribution is zero for every subset receives a value of zero. These axioms collectively guarantee a notion of fairness that LOO scores cannot provide: the efficiency axiom ensures complete credit allocation, and the symmetry axiom ensures that identical contributions are treated identically regardless of training randomness.

Unlike the LOO influence (Definition 2.1), which measures the loss change at a single specific subset ($\mathcal{D} \setminus \{z_i\}$), Shapley value averages marginal contributions across all 2^{N-1} subsets that exclude z_i, thereby capturing how a data point's contribution varies depending on which other data points are present. By considering all possible data compositions, Shapley values inherently account for the synergistic and antagonistic interactions that LOO misses. However, the resulting scores depend on the choice of utility function $v(\cdot)$, which is a design decision, and inherit the scale and units of $v(\cdot)$ rather than those of the loss function ℓ. The Shapley value provides the expected contribution of a data point to the learning algorithm across all hypothetical data compositions but does not estimate its loss change in any specific training process. This distinction is important for machine unlearning: Shapley values can identify which data points to prioritize for removal but do not directly predict the effect that removing a specific data point will have on the current model.

Alternative Game-Theoretic Solutions.

The Shapley value is not the only solution concept from cooperative game theory applicable to data valuation. Alternative *power indices* from the game theory literature offer different axiomatic tradeoffs that can be advantageous in specific settings, and several have been adapted to the data valuation context.

The *Banzhaf value* [25], originally developed for measuring voting power in weighted voting games, provides the most prominent alternative. Data Banzhaf [11] applies this concept to data valuation by replacing the size-dependent weighting in Eq. (2.7) with a uniform average over all subsets:

$$\beta_i(v) = \frac{1}{2^{N-1}} \sum_{S \subseteq \mathcal{D} \setminus \{z_i\}} \big[v(S \cup \{z_i\}) - v(S)\big]. \tag{2.8}$$

The key difference from the Shapley value is structural: while the Shapley value applies non-uniform weights that depend on coalition size $|S|$ (giving equal weight to each *permutation*), the Banzhaf value assigns equal weight to each *subset* regardless of its size. This uniform weighting sacrifices the efficiency axiom ($\sum_i \beta_i \neq v(\mathcal{D}) - v(\emptyset)$ in general), meaning that Banzhaf values do not fully allocate the total model utility. However, Wang and Jia [11] demonstrated that this tradeoff yields a substantial practical advantage: the Banzhaf value achieves a *Maximum Sample Complexity* (MSC) that is exponentially lower than that of the Shapley value, meaning it requires far fewer utility evaluations to estimate to a given accuracy. This lower sample complexity makes the Banzhaf value significantly more practical for large-scale data valuation.

Weighted Banzhaf Value [26] further extends the Banzhaf framework by introducing non-uniform weights over subsets, designed to improve robustness against noisy utility evaluations. In practice, the utility function $v(S)$ is estimated by training a model on subset S and evaluating it on a validation set, and the resulting estimates are inherently noisy due to the randomness of SGD. Li and Yu [26] showed that the weighted Banzhaf value can be configured to minimize the variance of the estimated data values under such noise, yielding more stable and reproducible data rankings than either Shapley or unweighted Banzhaf values.

Beta Shapley [27] takes a different approach by generalizing the Shapley value itself. It replaces the uniform weighting over coalition sizes in Eq. (2.7) with a Beta-distribution weighting Beta(α, β), yielding a tunable family of valuation scores parameterized by (α, β). When $\alpha = \beta = 1$, Beta Shapley reduces to the standard Shapley value; other choices allow practitioners to emphasize contributions in small coalitions (relevant for detecting individually impactful data points) or large coalitions (relevant for assessing marginal contributions in realistic training regimes). This flexibility makes Beta Shapley a unified framework that subsumes both the standard Shapley value and, as a limiting case, the Banzhaf value.

These alternative solution concepts highlight that the choice of game-theoretic foundation involves meaningful tradeoffs. The Shapley value provides the strongest fairness axioms (including efficiency) but is the hardest to estimate. The Banzhaf value sacrifices efficiency for dramatically lower estimation complexity and greater robustness to noise. Beta Shapley and weighted Banzhaf values offer tunable interpolations between these extremes, allowing practitioners to balance fairness, robustness, and computational cost according to application requirements.

Approximation Algorithms.

Regardless of the specific solution concept, all game-theoretic valuation methods face a shared computational bottleneck: evaluating the utility function $v(\cdot)$ on a combinatorial number of data subsets, with each evaluation requiring a full model training run. For the Shapley value, Eq. (2.7) involves $O(2^N)$ subset evaluations; for the Banzhaf value, Eq. (2.8) involves equally many. This exponential cost has motivated a substantial body of work on efficient approximation algorithms.

Data Shapley [1] employs Monte Carlo sampling of random permutations to estimate the Shapley value, converging to the true value as the number of sampled permutations increases but requiring multiple model retrainings per permutation. The same work introduced Gradient-based Shapley (G-Shapley) [1], which replaces full retraining with gradient-based utility estimation but still requires the model to reach convergence for accurate scoring. K-Nearest Neighbor Shapley (KNN-Shapley) [2] circumvents the retraining bottleneck entirely by restricting the model class to K-Nearest Neighbor (K-NN) classifiers, for which the Shapley value admits an exact polynomial-time computation; however, this restriction limits the method's applicability to models that are not well approximated by K-NN classifiers. One-For-All (OFA) [28] improves the convergence rate of Monte Carlo Shapley estimation through an optimized sample weighting vector. Other techniques [29–31] propose further computational shortcuts, though all remain constrained by the need for multiple model evaluations.

Approaches that avoid multiple subset evaluations and retraining have also emerged. Data Out-Of-Bag (Data-OOB) [32] eliminates retraining entirely by reusing out-of-bag predictions from bagging-based ensemble models, but this efficiency comes at the cost of restricting applicability to ensemble methods and introducing dependence on ensemble quality. In-Run Data Shapley [33] estimates data contributions during a single training run through Taylor approximations of the utility function, avoiding the need for retraining but inheriting sensitivity to the training order from the underlying gradient-based approximation. Class-wise Dhapley (CS-Shapley) [34] introduces a class-conditional variant that computes Shapley values separately within each class, improving performance on class-imbalanced datasets. Learning Algorithm Valuation (LAVA) [35] employs a dual formulation based on optimal transport to estimate data values without retraining, achieving computational efficiency at the cost of additional distributional assumptions.

Extensions and Applications.

The game-theoretic valuation framework has been adapted to contexts beyond standard data valuation. Active Data Shapley [36] integrates Shapley-based data valuation into the batch active learning loop, using data values to select informative data points for labeling. Domain-specific adaptations [37] extend the Shapley framework to federated learning, where the cooperative game structure naturally models the contributions of distributed data owners. Weighted SHapley Additive exPlanations (WeightedSHAP) [38] applies the Shapley approach to the distinct problem of explaining feature importance rather than valuing training data points, generalizing SHAP [39] by introducing non-uniform feature weighting. OpenDataVal [40] provides a standardized benchmark for comparing data valuation methods, including multiple game-theoretic algorithms, across consistent datasets and evaluation metrics.

Training Trajectory Methods

Training trajectory methods measure data influence by analyzing how individual data points shape the model parameters during the optimization process, rather than relying on the final converged model alone. This family of methods connects data influence to the optimization trajectory formalized in Eq. (1.7): each data point z_i affects the model through every iteration in which it participates in a mini-batch update, and training trajectory methods seek to quantify these per-iteration contributions and aggregate them across the training process. Unlike IFs, which require second-order curvature information, training trajectory methods operate with first-order gradient information, making them computationally more accessible for large-scale deep networks. Their influence scores are gradient-derived heuristics that correlate with data utility rather than estimates of the LOO loss change.

Gradient Inner Product Methods.

Tracing with Gradient Descent Influence (TracIn) [4] measures the influence of a training data point on a test data point by accumulating the reduction in test loss attributable to the training data point across all iterations in which it participates. The idealized form of TracIn decomposes the total change in test loss into per-data-point contributions using the gradient inner product:

$$\mathrm{TracIn}(z_i, z_{\text{test}}) = \sum_{t:\, z_i \in \mathcal{B}_t} \eta_t \, \nabla_\theta \ell\big(h_{\theta_t}(\mathbf{x}_i), y_i\big)^\top \, \nabla_\theta \ell\big(h_{\theta_t}(\mathbf{x}_{\text{test}}), y_{\text{test}}\big), \tag{2.9}$$

where the sum is taken over all iterations t in which z_i appears in the mini-batch $\mathcal{B}_t$, and η_t is the learning rate at iteration t. Intuitively, a positive gradient inner product at iteration t indicates that the gradient direction induced by z_i also reduces the loss on z_{test}, implying a beneficial influence.

In practice, storing the per-iteration gradients for all T iterations is infeasible. Pruthi et al. [4] proposed a computationally tractable approximation, TracIn-CP (CheckPoint-based TracIn), which evaluates the gradient inner product only at a small number of saved checkpoints $\{t_1, t_2, \ldots, t_K\}$:

$$\mathrm{TracIn\text{-}CP}(z_i, z_{\text{test}}) = \sum_{k=1}^{K} \eta_{t_k} \, \nabla_\theta \ell\big(h_{\theta_{t_k}}(\mathbf{x}_i), y_i\big)^\top \, \nabla_\theta \ell\big(h_{\theta_{t_k}}(\mathbf{x}_{\text{test}}), y_{\text{test}}\big). \tag{2.10}$$

This approximation reduces the computational cost from $O(T \cdot p)$ to $O(K \cdot p)$ per training-test data point pair, where $K \ll T$ is the number of checkpoints. TracIn has a natural connection to IFs: when evaluated at a single checkpoint corresponding to the final model and with the gradient inner product replaced by the inverse-Hessian-weighted inner product, TracIn recovers the IF estimate of Eq. (2.6) [4]. However, by operating across multiple points in the training trajectory rather than at the converged model alone, TracIn captures a broader view of data influence that partially mitigates the sensitivity to local curvature that limits IFs.

Gradient Scoring Methods.

A related family of methods assigns scalar influence scores to individual data points based on properties of their gradient vectors during training, without requiring a specific test data point as a reference. Gradient Normed (GraNd) [3] scores each data point by the expected norm of its loss gradient across training:

$$\mathrm{GraNd}(z_i) = \mathbb{E}_t\Big[\,\big\|\nabla_\theta \ell\big(h_{\theta_t}(\mathbf{x}_i), y_i\big)\big\|\,\Big], \tag{2.11}$$

where the expectation is taken over training iterations (or, in practice, over a set of early-training checkpoints). Data points with large gradient norms are interpreted as "difficult" or influential, as they induce large parameter updates. The same work introduced Error L_2-Norm (EL2N) [3], which simplifies the gradient norm computation by scoring data points using the L_2 norm of the error vector:

$$\mathrm{EL2N}(z_i) = \mathbb{E}_t\Big[\,\big\|h_{\theta_t}(\mathbf{x}_i) - \mathbf{e}_{y_i}\big\|_2\Big], \tag{2.12}$$

where $\mathbf{e}_{y_i}$ is the one-hot encoding of the label y_i. EL2N avoids the need for backpropagation entirely, requiring only a forward pass to compute the prediction error. Paul et al. [3] demonstrated that both GraNd and EL2N scores, even when computed using only a few early-training epochs, are effective for identifying data points that can be pruned from the training set without significant performance degradation.

Momentum-based Data Value Sourcing (MOSO) [41] extends the gradient-based approach by incorporating momentum, computing influence scores using the inner product between each data point's gradient and the momentum-averaged gradient direction. You Only Compress Once (YOCO) [42] builds upon the EL2N framework with a balanced dataset construction strategy that accounts for class distribution during data pruning. Capturing Training Dynamics (CTD) [43] formalizes how a data point's position in the optimization sequence affects its measured influence, quantifying trajectory-specific LOO influence estimation and revealing that data points encountered in earlier epochs receive systematically higher influence scores due to the larger gradient magnitudes characteristic of the early training phase. This training order bias is a pervasive issue across all gradient-based scoring methods: because gradient magnitudes decrease as the model converges, data points processed in the initial epochs accumulate disproportionate influence scores regardless of their intrinsic importance.

Gradient Alignment Methods.

Gradient alignment methods evaluate the influence of a data point by measuring the directional agreement between its gradient and a reference gradient, typically the average gradient over the dataset or a validation set. Cosine Gradient Similarity Value (CGSV) [44] computes the cosine similarity between a data point's gradient and the average gradient across all data points at each iteration, assigning higher value to data points whose gradients align with the overall optimization direction. Similar cosine-similarity-based approaches have been applied in federated learning [45, 46] to detect adversarial participants whose gradient updates diverge from the aggregate, and in data selection [47] to identify data subsets whose gradient structure aligns with a target distribution.

A fundamental limitation of gradient alignment methods is their inherently *local* perspective: they evaluate influence at individual iterations without a mechanism to assess how local gradient contributions accumulate into long-term effects on model convergence and generalization [44]. A data point whose gradient aligns poorly with the optimization direction at a single iteration may nonetheless contribute positively to overall model performance through its cumulative effect across many epochs, and vice versa. This limitation motivates the development of influence measurement frameworks that capture the dynamic, time-varying nature of data influence across the full training trajectory, as pursued in Chaps. 3 and 4.

Surrogate and Representation-Based Methods

Surrogate and representation-based methods approach data influence from an empirical modeling perspective: rather than deriving influence analytically or tracking gradient contributions, they learn the mapping from data composition to model output through auxiliary models or feature-space analysis. These methods incur upfront computational costs for training auxiliary models but offer amortized efficiency when evaluating influence across many data points or test inputs. Like the game-theoretic and gradient-based methods above, they produce custom influence scores tailored to specific downstream applications rather than estimates of the LOO loss change.

Surrogate Model Methods.

The Datamodels framework [5] learns a linear surrogate model that predicts the output of the trained model on a given test data point as a function of which training data points were included in the training set. Specifically, for a fixed test data point z_{test}, the datamodel fits coefficients $\boldsymbol{\alpha} \in \mathbb{R}^N$ such that:

$$f_{\text{DM}}(S;\ z_{\text{test}}) = \sum_{j=1}^{N} \alpha_j \cdot \mathbf{1}[z_j \in S] \ \approx\ g\big(h_{\mathcal{A}(S)}(\mathbf{x}_{\text{test}})\big), \tag{2.13}$$

where $S \subseteq \mathcal{D}$ is a random training subset, $\mathcal{A}(S)$ denotes the parameters obtained by training on S, and $g(\cdot)$ is a scalar summary of the model's output (e.g., the predicted probability of the correct class). The coefficients $\boldsymbol{\alpha}$ are estimated by training many models on independently sampled subsets $S \sim \text{Bernoulli}(\rho)^N$ (where each data point is included independently with probability ρ) and fitting a linear regression. The resulting coefficient α_j directly estimates the marginal influence of data point z_j on the test prediction, averaged over random subsets. Ilyas et al. [5] demonstrated that datamodels achieve surprisingly high predictive accuracy, explaining a large fraction of the variance in model outputs across training subsets. However, the approach requires training K independent models (typically K ranging from hundreds to thousands), yielding a total cost of $O(K \cdot E \cdot N \cdot p)$ that is substantial in absolute terms, though amortized across many influence queries.

TRaining data Attribution using Kernels (TRAK) [6] provides a scalable alternative by combining random projections with a kernel-based reformulation of the IF. TRAK projects the high-dimensional per-data-point gradients onto a low-dimensional random subspace, reducing the effective parameter dimension from p to a projection dimension $d_{\text{proj}} \ll p$. In this projected space, the influence of each data point on a test output is estimated via a ridge-regression formulation analogous to the Nadaraya-Watson kernel estimator, using a small ensemble of independently trained models for variance reduction. TRAK achieves substantially higher attribution accuracy than standard IFs (as measured by the linear datamodeling score introduced by Ilyas et al. [5]) while requiring far fewer model retrainings than the full Datamodels framework. Park et al. [6] demonstrated TRAK's applicability to large-scale image classifiers (ResNet, CLIP) and language models, establishing it as one of the most effective scalable attribution methods available.

Other proxy-based approaches include the data influence models of Yu et al. [48], which train auxiliary models to predict the influence of training data points on target tasks, and the Response-Conditioned Kernel-based Testing (RCKT) approach of Cui et al. [49], which computes proxy influence scores from the prediction changes induced by response flipping on individual data points.

Representation-Based Methods.

Representer Point Selection [50] provides a decomposition of the model's prediction at a test data point into a weighted sum of contributions from individual training data points, computed in the feature space of the model's penultimate layer. For a model with a linear final layer, the prediction at z_{test} can be written as:

$$h_{\theta^*}(\mathbf{x}_{\text{test}}) = \sum_{i=1}^{N} \alpha_i \, \mathbf{k}(\mathbf{x}_i, \mathbf{x}_{\text{test}}), \tag{2.14}$$

where $\mathbf{k}(\mathbf{x}_i, \mathbf{x}_{\text{test}}) = \boldsymbol{\Phi}(\mathbf{x}_i)^\top \boldsymbol{\Phi}(\mathbf{x}_{\text{test}})$ is the inner product in the learned feature space $\boldsymbol{\Phi}$, and the representer weights α_i are derived from the Karush-Kuhn-Tucker (KKT) conditions of the regularized training objective. The weight α_i quantifies the influence of training data point z_i on the test prediction through the similarity of their learned representations. This approach provides an interpretable decomposition but is limited by the assumption of a linear final layer and by the fact that the feature representations $\boldsymbol{\Phi}$ are themselves shaped by all training data points, complicating the attribution of influence to individual data points.

Summary of Characteristics.

The four families of methods surveyed above approach data influence from complementary perspectives, differing not only in their computational strategies but also in the application-driven motivations that shape their score definitions. Table 2.1 summarizes their key characteristics.

The fundamental divide is between LOO approximation methods and custom influence scoring methods. LOO retraining and IFs both target the actual loss change from removing a single data point: the former measures this quantity exactly at prohibitive cost, while the latter approximates it analytically under assumptions (convexity, convergence) that are rarely satisfied for deep networks. The computational and theoretical limitations of these LOO-based approaches have driven the development of the remaining three families, each of which designs its influence metric around a distinct practical objective rather than attempting to estimate the LOO loss change. Game-theoretic valuation methods define scores through axiomatic fairness guarantees, making them the principled choice for equitable credit allocation in data marketplaces and collaborative learning. Training trajectory methods define scores through gradient-derived properties (magnitudes, norms, directional alignments) that directly reflect how each data point drives parameter updates, making them natural tools for data cleansing and pruning. Surrogate methods learn explicit data-to-output mappings through auxiliary models, enabling efficient "which training data points caused this prediction?" queries that support model debugging and auditing at scale.

A common thread across all four families is the tension between computational tractability and the faithfulness of the influence estimate, a fundamental tradeoff that the methods of Chaps. 3 and 4 aim to navigate. The systematic limitations of these classical methods are analyzed in section "Limitations of Classical Methods".

Limitations of Classical Methods

The four families of classical data influence methods surveyed in section "Classical Data Influence Analysis" share a set of fundamental limitations that constrain their applicability to modern deep learning systems and, crucially, to machine unlearning. Rather than recapitulating the per-method weaknesses noted in each subsubsection above, this subsection organizes the limitations into five cross-cutting themes that collectively motivate the influence measurement frameworks developed in Chaps. 3 and 4.

Computational Scalability

The most immediate barrier to applying classical influence methods in practice is their computational cost. LOO retraining requires $O(N \cdot C_{\text{train}})$ total computation to evaluate all N training data points (Definition 2.1), where each of the N retraining runs costs $C_{\text{train}} = O(E \cdot N \cdot p)$. For a dataset of $N = 10^6$ data points and a model with $p = 10^9$ parameters trained over $E = 100$ epochs, a single LOO evaluation costs as much as the original training run, and evaluating the entire dataset multiplies this by N. Exact Shapley value computation (Definition 2.3) is exponentially worse, requiring $O(2^N)$ subset evaluations, each involving a full training run.

Approximation methods reduce these costs but introduce new scalability tradeoffs. Monte Carlo Shapley estimation [1] requires hundreds to thousands of model retrainings to achieve adequate convergence. Influence functions replace retraining

Table 2.1 Summary of classical data influence analysis methods. Cost is expressed relative to the full training cost $C_{\text{train}} = O(E \cdot N \cdot p)$

Method family	Primary motivation	Key assumptions	Per-point cost	Influence target	Deep learning applicability
LOO retraining	Exact LOO loss change (gold standard)	None (model-agnostic)	$O(C_{\text{train}})$	Test loss or global accuracy	Universal but prohibitive
Influence functions	Analytical LOO approximation	Convexity, convergence, twice-diff. loss	$O(p)$ per HVP iter.	Test data point loss	Limited; degrades in non-convex settings
Data shapley	Fair credit allocation (data markets, collaborative learning)	Utility function choice	$O(2^N \cdot C_{\text{train}})$	Utility $v(\cdot)$	Via approximations only
TracIn	Data cleansing & attribution via gradient tracking	Checkpoint availability	$O(K \cdot p)$	Test data point loss	Practical with checkpoints
GraNd/EL2N	Data pruning via gradient-based quality scoring	Training dynamics access	$O(p)$ per ckpt.	Self-influence (no test ref.)	Practical; forward-pass only for EL2N
CGSV	Data quality via gradient alignment detection	Reference gradient direction	$O(p)$ per iter.	Directional agreement	Practical but local
Datamodels/TRAK	Scalable attribution for model debugging & auditing	Approx. linear data-output map	$O(K \cdot C_{\text{train}})$ total	Test output prediction	After K model trainings
Representer points	Interpretable prediction decomposition	Linear output layer, L_2 reg.	$O(N \cdot d_{\text{feat}})$	Test prediction decomposition	Limited to linear output layer

with inverse-Hessian-vector products at $O(p)$ per iteration [9], but even with efficient approximations such as LiSSA [14], Arnoldi iteration [17], or EK-FAC [18], the per-query cost remains substantial for models with $p > 10^8$. Training trajectory methods such as TracIn [4] require storing and processing gradients at multiple checkpoints, with memory and computation scaling as $O(K \cdot p \cdot N)$ for K checkpoints across all N data points. Surrogate methods such as Datamodels [5] and TRAK [6] amortize costs by training K auxiliary models upfront, but the initial investment of $O(K \cdot C_{\text{train}})$ remains prohibitive for the largest models.

The scalability challenge is particularly acute for machine unlearning, where influence estimates may need to be computed *on demand* in response to deletion requests. A deployed system that receives deletion requests at irregular intervals cannot afford to wait for days of computation before executing each unlearning operation. This operational requirement for low-latency influence estimation distinguishes the unlearning context from offline data valuation tasks, where batch computation over extended time periods is acceptable. The real-time influence measurement framework of Chap. 3 is designed to address precisely this constraint, maintaining continuously updated influence estimates that can be queried at negligible marginal cost.

Violated Assumptions in Deep Learning

Influence functions, the most widely adopted analytical method for data influence estimation, rely on three assumptions that are systematically violated by modern deep neural networks: strict convexity, convergence to the global optimum, and twice-differentiable loss [9, 13].

Deep neural networks have non-convex loss landscapes characterized by multiple local minima, saddle points, and flat regions [51, 52]. The non-convexity invalidates the positive-definiteness assumption on the Hessian $\mathbf{H}_{\theta^*}$ (Eq. (2.5)), meaning the Hessian may be singular or indefinite, making the inverse-Hessian-vector product in Eq. (2.6) ill-defined without damping regularization. Standard practice adds a damping term $\lambda\mathbf{I}$ to the Hessian, computing $(\mathbf{H}_{\theta^*} + \lambda\mathbf{I})^{-1}$ in place of $\mathbf{H}_{\theta^*}^{-1}$, but Choe et al. [23] showed that the resulting influence estimates are highly sensitive to the choice of λ, and no principled method exists for selecting it. Furthermore, practical training with SGD (Definition 1.4) rarely converges to a global minimizer, especially under early stopping or learning rate scheduling. The residual gradient $\nabla_\theta \hat{R}(\theta^*; \mathcal{D})$ is non-zero at the final iterate, violating the first-order optimality condition upon which the derivation in Eq. (2.3) depends. Schioppa et al. [22] provided theoretical bounds on the approximation error under incomplete convergence, demonstrating that the error can grow unboundedly as the optimality gap increases.

Retraining-based methods (LOO, Shapley) do not require these analytical assumptions, but they are affected by a different form of assumption violation: the assumption of *deterministic* model training. In practice, stochastic training with random mini-batch ordering, random initialization, and dropout produces different models across independent runs on the same data. Wang et al. [11], Li et al. [12], and Sim et al. [10] demonstrated that both LOO and Shapley value rankings are sensitive to this training randomness, producing inconsistent data value orderings across runs. This instability undermines the reliability of influence estimates and raises the question of whether the influence of a data point is a well-defined quantity in the stochastic training regime, or whether it is inherently a distributional property that depends on the random seed.

Static Versus Dynamic Perspective

A fundamental conceptual limitation shared by influence functions, Shapley values, and representer point methods is their *static* perspective on data influence: they estimate influence at a single snapshot of the model, typically the final converged parameters $\boldsymbol{\theta}^*$, without accounting for how the influence of a data point evolves over the course of training. This static view conflicts with the dynamic reality of SGD-based optimization, in which the influence of each data point is inscribed into the parameters incrementally through a sequence of mini-batch updates (Eq. (1.7)).

Influence functions evaluate the effect of an infinitesimal perturbation at the converged model (Eq. (2.4)), capturing only the *local* sensitivity of the final parameters to a data point's contribution. They cannot distinguish between a data point that exerted strong influence in the early epochs but negligible influence later, and a data point that had moderate but consistent influence throughout training. This distinction is critical for machine unlearning, where the goal is to reverse the *cumulative* effect of a data point across the entire training trajectory, not merely to undo its marginal effect at convergence. Bae et al. [20] formalized this limitation through the concepts of *proximal mismatch* (influence functions approximate an infinitesimal perturbation, while data removal is a finite discrete operation) and *warm-start bias* (practical LOO experiments initialize retraining from the original parameters rather than from scratch, causing the retrained model to converge to a nearby local minimum that may not represent the true LOO solution).

Training trajectory methods such as TracIn (Eq. (2.9)) partially address the static limitation by accumulating gradient inner products across multiple training iterations. However, TracIn measures influence through *first-order* gradient projections and cannot capture the higher-order effects through which early-training influence compounds nonlinearly through subsequent parameter updates. Moreover, the checkpoint-based approximation TracIn-CP (Eq. (2.10)) samples influence at discrete points, potentially missing rapid changes in influence between checkpoints. Gradient scoring methods such as GraNd (Eq. (2.11)) and EL2N (Eq. (2.12)) provide even coarser temporal aggregation, averaging gradient norms over a small set of early epochs without tracking how the scores evolve as training progresses.

The need for a truly *dynamic* perspective that tracks data influence continuously across the training trajectory, capturing the time-varying nature of influence accumulation and accounting for the nonlinear interactions between consecutive updates, is a central motivation for the time-varying influence measurement framework of Chap. 4.

Training Order Bias and Inter-Data-Point Interactions

Two related limitations affect gradient-based influence methods: sensitivity to the order in which data points are processed during training, and the inability to capture interactions between data points.

Training Order Bias.

Gradient-based methods, including TracIn, GraNd, EL2N, and CGSV, assign influence scores that are systematically biased by the data point's position in the training sequence. In the early phases of SGD, loss values are high and gradient magnitudes are correspondingly large (the model is far from convergence), so data points processed in early iterations receive disproportionately high influence scores regardless of their intrinsic importance [3, 43]. As training progresses, gradient magnitudes diminish and data points processed in later epochs accumulate smaller scores. This bias is particularly problematic when the training data is not randomly shuffled between epochs or when certain data points appear more frequently in mini-batches (e.g., due to oversampling in imbalanced datasets). CTD [43] formalized this phenomenon by quantifying how data position in the optimization sequence affects the trajectory-specific LOO influence estimate, demonstrating that the bias can reverse the relative ordering of data points compared to the true LOO ranking.

The training order bias has direct consequences for machine unlearning. If an unlearning algorithm uses gradient-based influence scores to identify which parameters were most affected by a target data point, the bias can cause the algorithm to overestimate the influence of data points encountered early in training and underestimate the influence of data points encountered later. This systematic error can lead to incomplete removal of the target data point's influence or, conversely, to excessive modification of parameters that also encode the influence of retain-set data points.

Inter-Data-Point Interactions.

LOO retraining and influence functions evaluate data points in isolation, measuring the effect of adding or removing a single data point while holding the rest of the dataset fixed. This marginal perspective cannot capture the *combinatorial* interactions that arise when multiple data points are jointly added or removed [10]. In practice, data points interact through the optimization process: the gradient contributed by z_i at iteration t affects the parameters $\boldsymbol{\theta}_t$, which in turn affect the gradient that z_j contributes at iteration $t + 1$, creating cascading dependencies that are invisible to point-wise influence measures. For machine unlearning, inter-data-point interactions are especially relevant when the forget set $\mathcal{D}_f$ contains multiple related data points (e.g., all records from a single user, or a cluster of poisoned data points). The total influence of $\mathcal{D}_f$ on the model may differ substantially from the sum of individual influences $\sum_{z_i \in \mathcal{D}_f} \mathcal{I}(z_i)$, and unlearning algorithms that treat each data point independently may fail to account for the synergistic or antagonistic effects within the forget set.

Shapley values, by construction, account for subset interactions through the marginal contribution over all possible coalitions (Eq. (2.7)), but at the cost of exponential computation. No existing classical method simultaneously captures inter-data-point interactions, operates at scale, and provides a dynamic trajectory-level perspective, leaving a fundamental gap that the frameworks of Chaps. 3 and 4 aim to fill.

Discrepancy Between Influence Estimation and True LOO

A growing body of empirical evidence has documented significant discrepancies between influence function estimates and the true LOO influence in deep learning settings, calling into question the reliability of influence functions as a proxy for data removal effects.

Basu et al. [13] demonstrated through controlled experiments that influence function estimates can diverge substantially from true LOO influence in non-convex models, with Spearman rank correlations as low as 0.2 between the two quantities on standard vision benchmarks. The degradation is most severe for data points with large influence (the very data points that matter most for unlearning), because large-influence data points induce large parameter changes upon removal, amplifying the nonlinear effects that the first-order Taylor approximation in Eq. (2.4) cannot capture. Hammoudeh and Lowd [19] provided a comprehensive empirical evaluation across multiple model architectures and datasets, finding that the rank correlation between influence function estimates and true LOO varies widely depending on the model, dataset, and hyperparameters, with no reliable method for predicting when the approximation will be accurate.

The practical implication for machine unlearning is that influence function estimates cannot be trusted as reliable guides for identifying the parameters most affected by a target data point, nor for estimating the magnitude of the required parameter correction. An unlearning algorithm that relies on inaccurate influence estimates may produce a model that appears to have forgotten the target data (as measured by the influence function) but has in fact retained significant residual influence that would be revealed by a true LOO comparison or by a membership inference attack (Definition 1.10). This reliability gap motivates the development of influence measurement methods that either bypass the Taylor approximation entirely (as in the real-time framework of Chap. 3) or that incorporate higher-order trajectory information to reduce the approximation error (as in the time-varying framework of Chap. 4).

Summary and Transition.

The five limitations analyzed above, computational scalability, violated assumptions, the static-versus-dynamic mismatch, training order bias combined with the inability to capture inter-data-point interactions, and the empirical discrepancy between influence estimates and true LOO, collectively establish that no existing classical method provides influence estimates that are simultaneously accurate, efficient, dynamic, and scalable for the demands of machine unlearning in modern deep learning systems.

Relationships Between Data Influence and Unlearning

Data influence analysis and machine unlearning are connected by two fundamental relationships. First, influence analysis identifies *which* data points need to be removed from a trained model. Second, accurate influence estimation enables direct parameter modification to *implement* the removal efficiently. This subsection formalizes each relationship and briefly discusses how influence analysis also supports the verification of unlearning completeness.

Identifying Data Points for Unlearning

In many unlearning scenarios, the forget set $\mathcal{D}_f$ is not specified by an external request but must be discovered through data auditing. The security scenario (section "Application Scenarios and Design Principles") requires identifying poisoned or backdoored data points, the maintenance scenario requires locating mislabeled or biased data points, and even in the privacy scenario, provenance tracking is needed to determine which data points belong to a requesting user. Data influence analysis provides the primary tool for this identification.

Influence functions [9] identify training data points responsible for specific model errors by measuring each data point's effect on a given test prediction (Eq. (2.6)). Data points with anomalously large positive influence on incorrect predictions are candidates for mislabeling or poisoning. TracIn [4] provides a complementary perspective by tracing the cumulative gradient-level contribution of each data point across the training trajectory (Eq. (2.9)), revealing data points that consistently push the model toward undesirable behavior. Gradient alignment methods such as CGSV [44] flag data points whose gradient directions systematically diverge from the population average, a signature of adversarial injection. More recent surrogate approaches such as TRAK [6] enable scalable attribution that can rank all training data points by their contribution to a target behavior.

The accuracy of the identified forget set directly determines the effectiveness of the subsequent unlearning operation. False negatives (undetected malicious data points remaining in the retain set) leave residual adversarial influence in the model, while false positives (benign data points incorrectly included in the forget set) cause unnecessary knowledge loss. The limitations of classical influence methods analyzed in section "Limitations of Classical Methods", particularly the LOO approximation discrepancy (section "Discrepancy Between Influence Estimation and True LOO") and training order bias (section "Training Order Bias and Inter-Data-Point Interactions"), directly affect the reliability of forget-set identification.

Influence-Based Parameter Modification for Unlearning

If the influence of a data point on the model parameters can be accurately estimated, then unlearning reduces to reversing that influence through direct parameter modification, avoiding the cost of full retraining. This connection between influence estimation and unlearning is most clearly illustrated by the Newton-step update derived from the influence function.

The influence function approximation (Eq. (2.4)) estimates the parameter change induced by removing a single data point z_i from the training set:

$$\boldsymbol{\theta}^* - \boldsymbol{\theta}^*_{-i} \approx \frac{1}{N} \mathbf{H}_{\boldsymbol{\theta}^*}^{-1} \nabla_{\boldsymbol{\theta}} \ell\big(h_{\boldsymbol{\theta}^*}(\mathbf{x}_i), y_i\big). \tag{2.15}$$

This directly yields a one-step unlearning update [53, 54]:

$$\boldsymbol{\theta}^u = \boldsymbol{\theta}^* - \frac{1}{N} \mathbf{H}_{\boldsymbol{\theta}^*}^{-1} \nabla_{\boldsymbol{\theta}} \ell\big(h_{\boldsymbol{\theta}^*}(\mathbf{x}_i), y_i\big), \tag{2.16}$$

which subtracts the estimated influence of z_i from the current parameters. For a forget set containing multiple data points, the update extends to:

$$\boldsymbol{\theta}^u = \boldsymbol{\theta}^* - \frac{1}{N} \mathbf{H}_{\boldsymbol{\theta}^*}^{-1} \sum_{z_i \in \mathcal{D}_f} \nabla_{\boldsymbol{\theta}} \ell\big(h_{\boldsymbol{\theta}^*}(\mathbf{x}_i), y_i\big), \tag{2.17}$$

under the assumption that individual influences are approximately additive. This Newton-step approach is among the earliest and most direct influence-based unlearning methods, and it encapsulates the core principle: *accurate influence measurement enables efficient unlearning*.

However, the quality of this parameter modification is entirely dependent on the accuracy of the underlying influence estimate. When the assumptions of the influence function are satisfied (convex loss, converged model), the Newton-step update provides an accurate approximation to the retrained model $\boldsymbol{\theta}^*_{-i}$. When these assumptions are violated, as is typical in deep learning (section "Violated Assumptions in Deep Learning"), the update can produce a model that retains significant residual influence from the forget set or that unnecessarily degrades performance on the retain set. The static nature of the influence function (section "Static Versus Dynamic Perspective") further limits this approach, as it captures only the influence at convergence and misses the cumulative trajectory-level effects that a full retraining would reverse.

This fundamental dependence of unlearning quality on influence estimation quality has two important implications. First, it motivates the development of more accurate influence measurement frameworks (Chaps. 3 and 4) that can operate reliably in the non-convex, non-converged regime of deep learning. Second, it motivates dedicated unlearning algorithms (Chaps. 5 through 6) that go beyond simple influence subtraction by incorporating additional mechanisms such as fine-tuning on the retain set, structural partitioning of the training process, or in-training influence tracking.

A natural extension of the two relationships above is verification: after unlearning is performed, influence analysis can assess whether the forget set's influence has been adequately removed. By recomputing the influence of each $z_i \in \mathcal{D}_f$ on the unlearned model $\boldsymbol{\theta}^u$, one can estimate the residual influence and compare it against the baseline of a model that was never trained on z_i. In practice, this verification is typically combined with membership inference attacks (Definition 1.10) and distributional comparisons against the retrained model, forming a multi-faceted evaluation framework. The formal metrics and benchmarks for evaluating unlearning completeness are developed in section "Evaluation of Machine Unlearning".

Summary and Transition.

Data influence analysis supports machine unlearning through identification (locating the data points whose influence must be removed), implementation (enabling direct parameter modification through influence reversal), and verification (assessing whether the removal was successful). The effectiveness of all three roles is bounded by the accuracy and efficiency of the underlying influence estimates, establishing a direct link between the limitations of classical methods (section "Limitations of Classical Methods") and the achievable quality of influence-based unlearning. Having established these foundational connections, the remainder of this chapter surveys the existing unlearning literature. Section "Exact Machine Unlearning" examines exact unlearning methods that guarantee complete influence removal, while section "Approximate Machine Unlearning" covers approximate methods that trade guarantee strength for computational efficiency.

Exact Machine Unlearning

Exact machine unlearning provides the strongest form of data removal guarantee: the unlearned model must be statistically indistinguishable from a model that was never trained on the forget set. Formally, an unlearning algorithm $\mathcal{U}$ satisfies the exact unlearning condition (Definition 1.18) if the distribution of $\mathcal{U}(\boldsymbol{\theta}^*, \mathcal{D}, \mathcal{D}_f)$ is identical to the distribution of $\mathcal{A}(\mathcal{D} \setminus \mathcal{D}_f)$, where the randomness is over any stochasticity in the training algorithm $\mathcal{A}$ and the unlearning algorithm $\mathcal{U}$. This distributional equivalence ensures that no statistical test, including membership inference attacks (Definition 1.10), can distinguish an unlearned model from a retrained model, providing the strongest possible privacy guarantee for the data subjects whose data points are removed.

The exactness requirement imposes significant computational constraints. Naïve retraining, the most direct approach to exact unlearning, achieves distributional equivalence by construction but incurs the full cost of training a new model from scratch. The central challenge of exact unlearning research is to achieve the same distributional guarantee at substantially lower computational cost. This section surveys the methods that have been developed to address this challenge. Section "Naïve Retraining and SISA" introduces naïve retraining as the baseline and presents SISA (Sharded, Isolated, Sliced, and Aggregated training), the foundational framework that reduces unlearning cost through strategic data partitioning. Section "Existing Exact Unlearning Methods" reviews subsequent exact unlearning methods that build upon or complement the SISA framework. Section "Advantages and Limitations" analyzes the advantages and inherent limitations of exact unlearning approaches, motivating the approximate methods covered in section "Approximate Machine Unlearning".

Naïve Retraining and SISA

Exact unlearning requires that the unlearned model be distributionally indistinguishable from a model that was never trained on the forget set (Definition 1.18). This subsection presents the two foundational approaches that achieve this guarantee: naïve retraining, which serves as the gold standard by discarding the model and training from scratch, and the SISA framework, which partitions the training process to localize data influence and reduce the retraining cost.

Naïve Retraining

Naïve retraining, also known as full retraining or retraining from scratch, establishes the gold standard for exact unlearning and is widely used as a baseline for evaluating unlearning techniques. Given a forget set $\mathcal{D}_f$, the method discards the current model $\boldsymbol{\theta}^*$ entirely and trains a fresh model on the retain set from scratch.

Definition 2.4 (*Naïve Retraining*) Given a training algorithm $\mathcal{A}(\cdot)$, a training dataset $\mathcal{D}$, and a forget set $\mathcal{D}_f \subset \mathcal{D}$, naïve retraining produces the unlearned model by retraining on the retain set $\mathcal{D}_r = \mathcal{D} \setminus \mathcal{D}_f$:

$$\boldsymbol{\theta}^*_{\text{retrain}} = \mathcal{A}(\mathcal{D} \setminus \mathcal{D}_f). \tag{2.18}$$

As illustrated in Fig. 2.1, naïve retraining first discards the original model and removes the forget set from the training dataset, then trains a new model from scratch on the remaining data. Because the retrained model has never observed any data point in $\mathcal{D}_f$, the distributional equivalence required by Definition 1.18 is trivially satisfied. This approach requires no modification to the training algorithm $\mathcal{A}$, no auxiliary data structures, and no assumptions about the model architecture or loss function, making it universally applicable.

The limitation of naïve retraining is purely computational. Each unlearning request incurs the full training cost $C_{\text{train}} = O(E \cdot |\mathcal{D}_r| \cdot p)$, which for large-scale models can require days or weeks of computation on specialized hardware. For a model trained on $N = 10^6$ data points with $p = 10^9$ parameters over $E = 100$ epochs, even a single retraining run may cost thousands of GPU-hours, and the cost is incurred in full for every request regardless of whether the forget set contains one data point or one thousand. In deployment scenarios where unlearning requests arrive continuously, such as GDPR-mandated deletion requests that must be honored within a bounded time window, naïve retraining is operationally infeasible. Furthermore, naïve retraining requires access to the full original training dataset $\mathcal{D}$, which limits its feasibility in settings where the original data are no longer available due to storage constraints, access restrictions, or prior deletion of other data points.

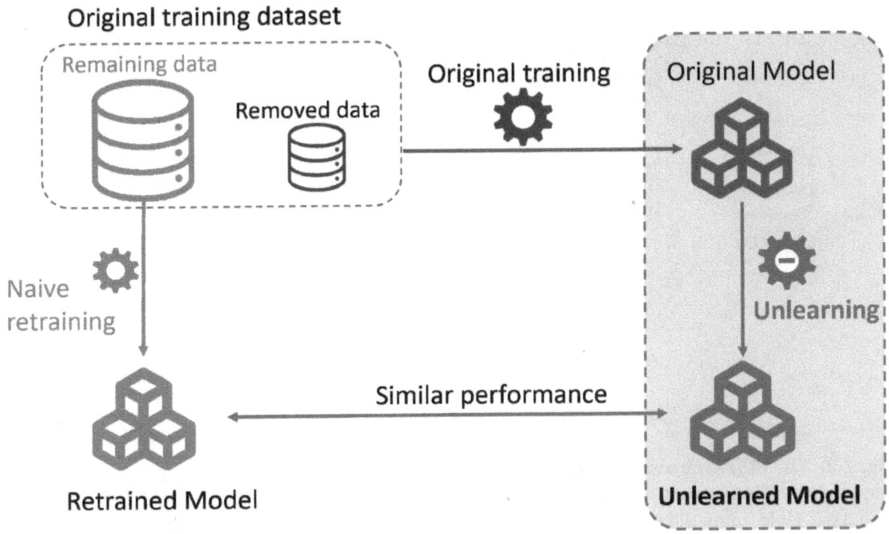

Fig. 2.1 Illustration of naïve retraining and machine unlearning. Naïve retraining discards the original model and trains from scratch on the retain set $\mathcal{D}_r = \mathcal{D} \setminus \mathcal{D}_f$, while machine unlearning methods aim to produce an equivalent result without full retraining

The SISA Framework

The Sharding, Isolation, Slicing, and Aggregation (SISA) framework [55] reduces the computational cost of exact unlearning by restructuring the training process so that each data point's influence is confined to a small, independently trained sub-model. When an unlearning request arrives, only the affected sub-model requires retraining, avoiding the full-dataset retraining cost of the naïve approach.

Framework Design.

As illustrated in Fig. 2.2, SISA organizes training through four complementary mechanisms. *Sharding* partitions the training dataset $\mathcal{D}$ into S disjoint subsets $\{\mathcal{D}^{(1)}, \mathcal{D}^{(2)}, \ldots, \mathcal{D}^{(S)}\}$, where $\bigcup_{s=1}^{S} \mathcal{D}^{(s)} = \mathcal{D}$ and $\mathcal{D}^{(s)} \cap \mathcal{D}^{(s')} = \emptyset$ for $s \neq s'$, with each shard containing approximately N/S data points. *Isolation* trains a separate constituent model $\boldsymbol{\theta}^{(s)} = \mathcal{A}(\mathcal{D}^{(s)})$ on each shard independently, ensuring that the influence of any data point $z_i \in \mathcal{D}^{(s)}$ is confined to the constituent model $\boldsymbol{\theta}^{(s)}$. *Slicing* further divides each shard into R ordered slices $\mathcal{D}^{(s)} = \mathcal{D}_1^{(s)} \cup \mathcal{D}_2^{(s)} \cup \cdots \cup \mathcal{D}_R^{(s)}$ and trains the constituent model incrementally on these slices, saving intermediate checkpoints $\boldsymbol{\theta}_r^{(s)}$ after incorporating each slice r. These checkpoints enable retraining to resume from the last clean state that does not include the data point to be unlearned, rather than retraining the constituent model from scratch. *Aggregation* combines the predictions of all S constituent models to form the final output, typically through majority voting for classification or averaging for regression, partially compensating for the reduced training data available to each individual model.

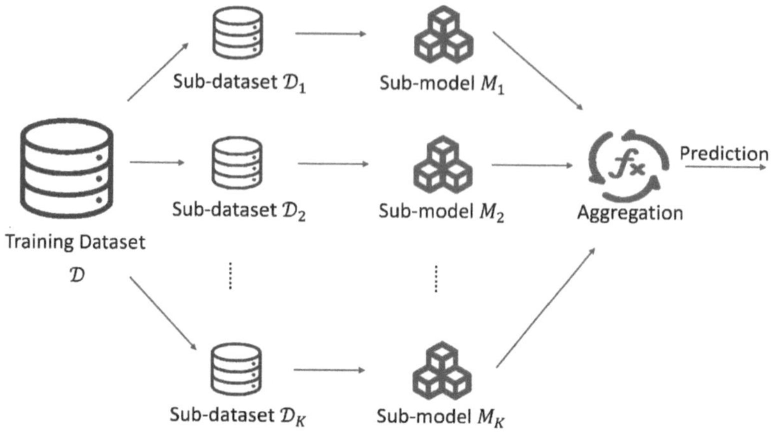

Fig. 2.2 The SISA framework. Training data are partitioned into S shards, each further divided into R slices. Constituent models are trained incrementally with checkpoints saved at each slice boundary. Unlearning a data point requires retraining only the affected constituent model from the appropriate checkpoint

Unlearning Procedure and Computational Cost.

When an unlearning request targets a data point $z_i \in \mathcal{D}_r^{(s)}$ (residing in slice r of shard s), SISA removes z_i from shard $\mathcal{D}^{(s)}$, restores the checkpoint $\boldsymbol{\theta}_{r-1}^{(s)}$ saved before slice r was incorporated, and retrains the constituent model from this checkpoint through the remaining slices $r, r+1, \ldots, R$ (now excluding z_i). Because the checkpoint $\boldsymbol{\theta}_{r-1}^{(s)}$ was produced without any exposure to z_i, and the subsequent retraining on the modified slices does not include z_i, the resulting constituent model satisfies the exact unlearning guarantee.

The initial training cost of SISA is comparable to naïve training. Each of the S constituent models is trained on approximately N/S data points, so the total training cost is:

$$C_{\text{train}}^{\text{SISA}} = S \cdot O\left(E \cdot \frac{N}{S} \cdot p\right) = O(E \cdot N \cdot p) = C_{\text{train}}. \tag{2.19}$$

The overhead lies in maintaining S separate models and $S \cdot R$ checkpoints, introducing storage costs of $O(S \cdot R \cdot p)$ parameters.

For a single-data-point unlearning request where $|\mathcal{D}_f| = 1$, exactly one shard is affected. If the data point resides in slice r of that shard, the retraining cost is:

$$C_{\text{unlearn}}^{\text{SISA}} = O\left(E \cdot \frac{N}{S} \cdot \frac{R-r+1}{R} \cdot p\right). \tag{2.20}$$

In expectation over uniformly distributed deletion requests, the expected cost is $O(E \cdot N \cdot p \,/\, (2S))$ when slicing is used optimally, representing a speedup factor of approximately $2S$ over naïve retraining. For $S = 20$ shards, this corresponds to a roughly $40\times$ reduction in expected unlearning cost.

Increasing the number of shards S reduces the per-request unlearning cost but also reduces the training data available to each constituent model from N to N/S, potentially degrading model accuracy. This tradeoff is governed by two opposing forces: smaller shards yield faster unlearning but provide each constituent model with less training data, while larger shards preserve model quality but diminish the computational advantage over naïve retraining.

Limitations.

Despite its computational advantages, SISA has several inherent limitations. The sharding strategy assumes that the training data can be meaningfully partitioned into independent subsets, an assumption that may not hold when the model must learn complex interactions between data points distributed across different shards. The isolation requirement prevents constituent models from sharing information during training, which can degrade performance on tasks where cross-shard data dependencies are important. The storage overhead of maintaining $S \cdot R$ intermediate checkpoints is substantial: for a model with $p = 10^9$ parameters, $S = 20$ shards, and $R = 10$ slices, the total checkpoint storage reaches $200 \times p \approx 800$ GB in single precision, posing practical challenges for deployment. Furthermore, when unlearning requests are concentrated in a single shard (a non-uniform deletion pattern), the expected speedup degrades because the same constituent model must be retrained repeatedly. These limitations have motivated the development of more sophisticated exact unlearning methods that reduce the dependence on data partitioning and lower storage costs, as discussed in section "Existing Exact Unlearning Methods".

Existing Exact Unlearning Methods

The SISA framework (section "The SISA Framework") established the core principle that exact unlearning can be made efficient by confining each data point's influence to a small, independently trained sub-model. Subsequent work has extended this principle along two axes. The first line of research adapts the SISA partitioning strategy to model families whose structure does not naturally decompose into independent shards, including tree-based ensembles, graph neural networks, clustering algorithms, and federated learning systems. The second line proposes structural improvements to SISA itself, addressing limitations in data partitioning, checkpoint management, and accuracy preservation. Beyond the SISA paradigm, a smaller body of work achieves exact unlearning through alternative mechanisms, such as intermediary representations that decouple the model from the raw training data or efficient retraining via curvature approximations. This subsection surveys both categories and concludes with a comparative analysis of their strengths and limitations.

Exact Unlearning for Tree-Based Models

Tree-based ensemble methods such as random forests exhibit a natural affinity with the SISA paradigm: each tree in the ensemble is trained on a different subset of the data, effectively serving as an independent shard whose retraining cost is modest relative to the full ensemble. Exact unlearning for random forests leverages this inherent modularity, retraining only the trees whose training subsets contain the data point to be forgotten while leaving the remaining trees unchanged. The key challenge is to reduce the per-tree retraining cost by exploiting the structure of decision tree construction.

DaRE (Data Removal-Enabled Forest) [56] achieves this by introducing a hybrid node structure within each tree. Nodes in the upper levels of the tree select split attributes and thresholds uniformly at random, making them largely independent of any individual data point. Nodes in the lower levels use greedy optimization of split criteria such as the Gini index or mutual information, where the split decision depends on the data. DaRE caches sufficient statistics at each greedy node and stores the training data points reaching each leaf, so that when an unlearning request arrives, only the affected subtrees rooted at greedy nodes need to be updated. This caching strategy avoids retraining the entire tree, reducing the per-tree unlearning cost to the depth of the affected subtree rather than the full tree height. However, the worst-case cost (when the data point affects the root split) remains comparable to full tree retraining, and the cached statistics impose additional storage overhead proportional to the number of greedy nodes.

HedgeCut [57] takes a complementary approach by precomputing alternative subtrees for splits that are sensitive to data removal. HedgeCut introduces the concept of *split robustness*: a split decision is robust if removing any single data point from the node's training set would not change the chosen split attribute or threshold. For non-robust splits, HedgeCut precomputes and stores *subtree variants* that correspond to alternative split decisions. When an unlearning request arrives, the method identifies any non-robust splits affected by the removed data point and replaces the corresponding subtrees with their precomputed variants, yielding an unlearning operation that requires no retraining and completes in time proportional to the tree depth. This constant-time unlearning comes at the cost of additional storage for the precomputed variants and depends on the accuracy of predicting which splits may be affected by future removal requests.

Exact Unlearning for Graph-Based Models

Graph-structured data poses a distinctive challenge for exact unlearning because the influence of any single node propagates through the graph's connectivity structure. In a graph neural network (GNN), each node's representation is computed by aggregating information from its neighbors, so removing a node affects not only its own representation but also those of all nodes within its multi-hop neighborhood. Naively applying SISA-style partitioning to graph data risks severing edges that carry important structural information, degrading the quality of the learned representations.

GraphEraser [58] addresses this challenge through a three-phase approach designed specifically for GNN unlearning. In the first phase, *balanced graph partitioning*, the graph is partitioned into S disjoint subgraphs using community detection algorithms that preserve intra-community edge density while balancing the partition sizes. This partitioning strategy aims to minimize the structural information lost by the sharding process. In the second phase, *shard model training*, an independent GNN sub-model is trained on each subgraph. In the third phase, *shard model aggregation*, a learning-based aggregation mechanism assigns importance weights to the sub-models, combining their predictions into a single output. When a node needs to be unlearned, GraphEraser removes the node from its subgraph, retrains only the affected sub-model, and updates the aggregation weights. The community-aware partitioning preserves more of the graph's structural information than random sharding, but the fundamental tradeoff between partition quality and unlearning efficiency persists: finer partitions enable faster unlearning but reduce the structural context available to each sub-model.

RecEraser [59] specializes the SISA framework for recommendation systems, where user-item interactions form a bipartite graph and collaborative filtering requires capturing cross-user and cross-item dependencies. RecEraser proposes three complementary partitioning strategies based on users, items, and interactions, respectively, each producing balanced shards that preserve different aspects of the collaborative structure. An attention-based adaptive aggregation mechanism learns to weight each sub-model's predictions based on both the local collaborative information captured within each shard and the global collaborative patterns across all shards. When an unlearning request targets a user-item interaction, only the affected sub-model and the aggregation layer require retraining. Compared to GraphEraser's static importance weighting, RecEraser's adaptive aggregation yields more accurate recommendations after unlearning by dynamically rebalancing the contribution of each sub-model.

Exact Unlearning for Clustering

DC-k-means [60] applies the SISA paradigm to k-means clustering through a hierarchical divide-and-conquer strategy. The training data are randomly partitioned into multiple subsets, each assigned to a leaf node in a perfect w-ary tree of height h. A local k-means model is fitted on each leaf subset, and the results are aggregated by recursively merging cluster centroids from the leaf nodes upward to the root. When a data point needs to be unlearned, the procedure locates the corresponding leaf node, updates the local k-means model to exclude the removed data point, and propagates the centroid changes upward through the tree. This hierarchical propagation avoids rerunning k-means on the full dataset, reducing the per-removal cost from $O(N)$ to $O(N/w^h)$ for the local update plus $O(h \cdot k)$ for the tree propagation, where k is the number of clusters. Ginart et al. [60] provided theoretical guarantees on both the removal efficiency and the clustering quality, showing that the hierarchical aggregation introduces negligible quality loss relative to standard k-means for moderate tree depths.

Exact Unlearning for Federated Learning

Federated learning introduces additional complexity for exact unlearning because training data are distributed across multiple clients and the global model is constructed through iterative aggregation of local updates rather than centralized training. When a client requests the removal of its data, the unlearning mechanism must ensure that the client's contributions to the global model are fully eliminated without requiring all clients to retrain from scratch.

KNOT [61] adapts the SISA framework to the federated setting by introducing a clustered aggregation mechanism that groups clients into multiple clusters. The server performs model aggregation only within each cluster, while different clusters train asynchronously. When a client requests unlearning, only the clients within the same cluster need to retrain, while other clusters continue training unaffected. To determine the optimal client-cluster assignment, KNOT formulates a lexicographic minimization problem that jointly considers training speed and model similarity between clients, solving it efficiently as a linear program. The asynchronous cluster structure confines the retraining cost to a single cluster (analogous to a single shard in SISA), but the quality of the global model depends on the cluster assignment, and non-uniform client distributions can lead to imbalanced clusters that degrade either unlearning efficiency or model accuracy.

Liu et al. [62] propose an alternative approach that avoids data partitioning entirely. When a client requests data removal, all clients perform local data removal, followed by an efficient retraining procedure that uses a first-order Taylor approximation based on the Quasi-Newton method combined with a low-cost Hessian approximation. This approach reduces the computational and communication costs of full retraining while maintaining model performance, but its accuracy depends on the quality of the Hessian approximation, which may degrade in highly non-convex settings.

Exact Unlearning for Large Vision Models

The scale of modern large vision models, often comprising hundreds of millions or billions of parameters, makes both naïve retraining and standard SISA partitioning impractical. LMEraser [63] addresses this by combining data partitioning with parameter-efficient fine-tuning techniques. Rather than training a full sub-model on each shard, LMEraser maintains a shared pretrained backbone and trains lightweight prompt-tuning modules on each data partition. When an unlearning request arrives, only the prompt-tuning module associated with the affected partition requires retraining, leaving the shared backbone unchanged. This approach dramatically reduces both the per-shard training cost and the storage overhead, making exact unlearning feasible for large vision models under the proposed visual prompt-tuning architecture at the cost of restricting the expressiveness of each sub-model to the capacity of the prompt-tuning module.

Advantages and Limitations

Advantages.

All methods share the defining advantage of exact unlearning: the unlearned model is distributionally equivalent to one trained from scratch on the retain set $\mathcal{D}_r = \mathcal{D} \setminus \mathcal{D}_f$ (Definition 1.18). This guarantee is unconditional in the sense that it holds for any model architecture, loss function, and optimization algorithm, requiring no assumptions about the loss landscape or training dynamics. For applications governed by strict regulatory requirements such as GDPR's right to erasure, this unconditional guarantee is the gold standard against which all other unlearning methods are measured. The domain-specific adaptations surveyed in section "Existing Exact Unlearning Methods" further demonstrate that exploiting model structure, whether the modularity of tree ensembles, the community structure of graphs, or the parameter efficiency of prompt tuning, can reduce unlearning costs by orders of magnitude relative to naïve retraining while preserving this guarantee in full.

Limitations.

These advantages are accompanied by several fundamental limitations that constrain practical applicability. First, all SISA-based methods incur substantial *storage overhead*: SISA itself maintains $S \cdot R$ checkpoints of $O(p)$ parameters each (Eq. (2.19)), DaRE and HedgeCut cache per-node statistics or precomputed subtree variants, and even the summation-based approach of Cao et al. [64] requires maintaining intermediary representations. For a model with $p = 10^9$ parameters, $S = 20$ shards, and $R = 10$ slices, the checkpoint storage alone reaches approximately 800 GB in single precision. Second, no existing method simultaneously achieves model agnostic applicability, low overhead, and efficient unlearning; each is either restricted to a specific model family (DaRE to tree ensembles, GraphEraser to GNNs, DC-k-means to clustering) or requires assumptions about data partitionability that may not hold when cross-shard dependencies are important. Third, SISA-style partitioning reduces the effective training data per constituent model from N to N/S, and while moderate sharding ($S \leq 20$) incurs less than 2% accuracy loss on standard benchmarks [55], the degradation can be severe for billion-parameter models that require massive unified corpora to achieve full capability. Fourth, existing methods focus on single-request or small-batch scenarios and lack formal guarantees on model utility after many sequential unlearning requests, a critical gap for deployment under continuous deletion demands. Fifth, both naïve retraining and SISA-based methods require access to the original training data $\mathcal{D}$, making exact unlearning infeasible when data are no longer available due to storage constraints or prior deletion. These limitations collectively motivate the development of approximate unlearning methods (section "Approximate Machine Unlearning"), which relax the distributional equivalence requirement in exchange for dramatically reduced computational costs and broader model applicability.

Approximate Machine Unlearning

Exact unlearning methods (section "Exact Machine Unlearning") guarantee that the unlearned model is distributionally equivalent to one retrained from scratch, but this guarantee requires retraining either the full model or constituent sub-models, incurring costs that scale with model size and dataset complexity. Approximate unlearning relaxes the distributional equivalence requirement of Definition 1.18, instead seeking to reduce the influence of the forget set $\mathcal{D}_f$ on the model to a level that is provably bounded or empirically negligible, without full retraining. This relaxation yields three practical advantages over exact methods. First, approximate methods achieve substantially lower computational costs because they modify existing model parameters rather than retraining from scratch, typically requiring only a small number of gradient steps or a single closed-form parameter update. Second, they incur minimal storage overhead because they operate on the existing model without maintaining multiple sub-models or checkpoints. Third, they offer greater flexibility across model architectures because they do not require the data partitioning or architectural modifications that exact methods demand. These advantages come at the cost of weaker removal guarantees: approximate methods do not ensure that the unlearned model is indistinguishable from a retrained model, and the residual influence of the forget set must be carefully bounded or empirically verified.

Approximate unlearning methods can be broadly organized into two families based on their mechanism for modifying the model. *Parameter manipulation methods* (section "Parameter Manipulation Methods") compute closed-form or single-step parameter updates derived from analytical formulas, such as influence functions or localized weight edits, and apply them directly to the model parameters without iterative optimization. *Fine-tuning and reweighting methods* (section "Fine-Tuning and Reweighting Methods") use iterative optimization to adjust the model, performing multiple gradient steps on objectives designed to suppress the influence of the forget set while preserving performance on the retain set. Section "Advantages and Limitations" provides a comparative analysis of the advantages and limitations of approximate methods as a whole.

Parameter Manipulation Methods

Parameter manipulation methods modify model parameters through closed-form updates or single-step corrections derived from analytical characterizations of data influence. These methods avoid iterative optimization entirely, instead computing the parameter adjustment needed to approximately remove the forget set's influence and applying it in a single operation. The resulting computational cost is typically dominated by a matrix inversion or a localized weight update, both of which are substantially cheaper than full retraining. This subsection surveys three principal approaches: influence function methods, which use curvature information to estimate the parameter change induced by data removal; knowledge editing methods, which target specific information encoded in individual model components; and structural modification methods, which alter the model architecture itself to eliminate data influence.

Influence Function Methods

Influence function methods apply the analytical framework developed in section "Leave-One-Out Retraining and Influence Functions" to the unlearning problem: given the influence function's estimate of how each training data point affects the model parameters, the method reverses this influence by applying a corrective parameter update. The key insight is that the Newton update derived from the influence function provides a first-order approximation to the parameter change that would result from removing a data point and retraining, enabling approximate unlearning without any retraining.

Certified Removal via Newton Updates.

Guo et al. [53] introduced the first certified removal mechanism for approximate unlearning, establishing the connection between influence functions and provable unlearning guarantees. Consider a model trained by minimizing the L_2-regularized empirical risk:

$$\hat{R}_\lambda(\boldsymbol{\theta};\, \mathcal{D}) = \sum_{z_i \in \mathcal{D}} \ell(z_i;\, \boldsymbol{\theta}) + \frac{\lambda N}{2} \|\boldsymbol{\theta}\|_2^2, \tag{2.21}$$

where $\ell(z_i;\, \boldsymbol{\theta})$ denotes the per-data-point convex loss and λ is the regularization strength. To protect gradient information from revealing properties of removed data points, the training process incorporates a random perturbation vector $\mathbf{b}$, yielding the perturbed objective $\hat{R}_{\lambda,\mathbf{b}}(\boldsymbol{\theta};\, \mathcal{D}) = \hat{R}_\lambda(\boldsymbol{\theta};\, \mathcal{D}) + \mathbf{b}^\top \boldsymbol{\theta}$, with optimal parameters $\boldsymbol{\theta}^* = \arg\min_{\boldsymbol{\theta}} \hat{R}_{\lambda,\mathbf{b}}(\boldsymbol{\theta};\, \mathcal{D})$.

When a data point z_i is to be removed, the Newton update removal mechanism proceeds in two steps. First, the loss gradient at z_i is computed as $\boldsymbol{\Delta} = \lambda\boldsymbol{\theta}^* + \nabla_{\boldsymbol{\theta}} \ell(z_i;\, \boldsymbol{\theta}^*)$. Second, the model parameters are adjusted using the inverse Hessian of the retain-set loss to produce the unlearned parameters:

$$\boldsymbol{\theta}^- = \boldsymbol{\theta}^* + \mathbf{H}_{\boldsymbol{\theta}^*}^{-1} \boldsymbol{\Delta}, \qquad \text{where} \quad \mathbf{H}_{\boldsymbol{\theta}^*} = \nabla_{\boldsymbol{\theta}}^2 \hat{R}_\lambda(\boldsymbol{\theta}^*;\, \mathcal{D}_r). \tag{2.22}$$

This one-step Newton update approximates the parameters that would be obtained by retraining on $\mathcal{D}_r = \mathcal{D} \setminus \{z_i\}$ (see the influence function in Eq. (2.4)). Combined with the random perturbation $\mathbf{b}$ added during training, this mechanism achieves (ε, δ)-certified removal: an adversary observing the unlearned model cannot distinguish it from a model retrained from scratch with probability exceeding the specified privacy budget. This certified guarantee holds rigorously for L_2-regularized models with convex losses but does not extend to the non-convex settings typical of deep neural networks.

Reducing Hessian Computation Costs.

The principal computational bottleneck in Eq. (2.22) is the Hessian inversion $\mathbf{H}_{\theta^*}^{-1}$, which costs $O(p^3)$ for p parameters and can be numerically unstable when the condition number of $\mathbf{H}_{\theta^*}$ is large. Subsequent work has proposed several strategies to mitigate this cost.

Sekhari et al. [65] avoid requiring the full training dataset during unlearning by precomputing and storing the aggregated Hessian statistic $\nabla_{\theta}^2 \hat{R}_{\lambda}(\theta^*; \mathcal{D})$ during training. At unlearning time, the retain-set Hessian is reconstructed by subtracting the forget set's contribution, and the parameter update is computed from this corrected Hessian. This approach reduces both storage requirements (only the compact Hessian statistic needs to be retained rather than the full dataset) and computation costs (a single Hessian subtraction replaces recomputation over the full dataset). The authors also emphasize the importance of bounding the test loss rather than just the training loss, and they introduce the notion of *deletion capacity*, which quantifies the maximum number of data points that can be sequentially removed before the unlearning approximation degrades beyond an acceptable threshold.

Suriyakumar et al. [66] propose a more computationally efficient algorithm for online data removal by replacing the per-request Hessian inversion with the *infinitesimal jackknife*, a technique that approximates the influence of excluding a data point by reusing a single precomputed Hessian inverse across multiple removal requests. This avoids the need to compute and invert a different Hessian matrix for each request, as was required by Guo et al. [53] and Sekhari et al. [65]. By integrating the infinitesimal jackknife with Newton methods, their algorithm also accommodates models trained with non-smooth regularizers such as the L_1 penalty, broadening applicability beyond the smooth L_2-regularized setting. Suriyakumar et al. further analyze certified unlearning under distribution shift, addressing the practically relevant scenario where the data distribution changes between training and unlearning time.

Mehta et al. [67] address the scalability of influence-based unlearning to deep neural networks, where the full Hessian is infeasible to compute or invert. They introduce L-CODEC, a selection scheme based on conditional independence that identifies a small subset of parameters whose updates are sufficient to approximately remove the forget set's influence. Building on this, they propose L-FOCI, which incrementally constructs a minimal set of influential parameters and applies a blockwise Newton update only to this subset. By restricting the Hessian inversion to a low-dimensional parameter subspace, this approach reduces the computational cost from $O(p^3)$ to $O(k^3)$ where $k \ll p$ is the number of selected parameters, making influence-based unlearning feasible for deep networks with millions of parameters.

Extensions to Comprehensive Removal and Fine-Grained Correction.

Several works extend the influence function framework beyond the original single-data-point, linear-model setting. PUMA [68] modifies all trainable parameters rather than only the final linear layer, providing more thorough data removal for deep networks. PUMA uses the influence function to measure each remaining data point's contribution to model performance and reweights these contributions to compensate for the removed data, prioritizing the preservation of model utility rather than solely minimizing the residual influence of the forget set.

Tanno et al. [69] propose a Bayesian continual learning approach that uses influence functions to identify the most detrimental training data points responsible for observed model failures, then erases their influence by approximating a counterfactual posterior distribution in which the harmful data points are assumed absent. Warnecke et al. [70] generalize the influence function framework to enable corrections at granularities finer than the data-point level, including the removal of specific features and labels. By reformulating influence estimation as a form of data modification, they derive closed-form parameter updates that can efficiently correct problematic features and labels even when large portions of the training data are affected.

Graph-Specific Influence Methods.

Graph-structured data poses a distinctive challenge for influence-based unlearning because the influence of any single node propagates through the graph's connectivity structure to its multi-hop neighbors (see the section "Exact Unlearning for Graph-Based Models"). Standard influence functions, which treat data points as independent, fail to capture these structural dependencies.

GIF (Graph Influence Function) [71] extends the classical influence function to graph neural networks by introducing an additional loss term that accounts for the influence of a removed node or edge on its neighbors. GIF estimates the parameter changes in response to ϵ-mass perturbations in the removed data while simultaneously adjusting for the cascading effects on neighboring node representations, providing a closed-form approximate unlearning solution that respects the graph structure. Chien et al. [72] derive theoretical guarantees for node, edge, and feature deletion specifically for simple graph convolutions and their generalized PageRank extensions, establishing certified removal bounds in the graph setting. GNNDelete [73] introduces a deletion operator that enforces two properties: *deleted edge consistency*, which ensures that removing an edge does not affect representations of other edges in the same neighborhood, and *neighborhood influence*, which confines the deletion effect to direct neighbors rather than the entire graph.

Knowledge Editing Methods

Knowledge editing methods target specific information encoded within model parameters, modifying localized components to remove or alter particular knowledge without global retraining. These methods are particularly relevant for large language models (LLMs), where the scale of the model makes both retraining-based exact methods and full influence function computation infeasible.

Task Arithmetic.

Task Vectors [74] provide a simple but effective mechanism for removing learned capabilities from a model. A task vector $\boldsymbol{\tau}$ is defined as the difference between the parameters of a model fine-tuned on a specific task and the parameters of the pretrained base model: $\boldsymbol{\tau} = \boldsymbol{\theta}_{\text{ft}} - \boldsymbol{\theta}_{\text{pre}}$. Subtracting a scaled task vector from the fine-tuned model, $\boldsymbol{\theta}^{-} = \boldsymbol{\theta}_{\text{ft}} - \alpha\boldsymbol{\tau}$, effectively removes the capability acquired during fine-tuning, where α controls the removal strength. This approach extends naturally to unlearning: by computing the task vector corresponding to training on the forget set and subtracting it, the model's knowledge of the forget set can be approximately suppressed. Task arithmetic is computationally inexpensive (requiring only vector subtraction) but provides no formal guarantees on removal completeness, and the linear assumption underlying task vector composition may not hold when the forget set's influence is distributed nonlinearly across the parameters.

Localized Parameter Editing.

A complementary line of work identifies and modifies specific model components, such as individual layers, neurons, or attention heads, that are responsible for encoding target knowledge. ROME (Rank-One Model Editing) [75] demonstrates precise fact editing in transformer-based language models by applying targeted rank-one updates to the weight matrices of specific feed-forward layers, effectively overwriting stored factual associations. MEMIT (Mass-Editing Memory in a Transformer) [76] extends ROME to simultaneously edit multiple facts by distributing rank-one updates across several layers, enabling batch knowledge removal. Knowledge Neurons [77] identifies neurons that are activated specifically in response to particular factual knowledge, enabling selective suppression through targeted neuron modification. DEPN [78] provides a method for detecting and editing prediction-relevant neurons to remove specific knowledge while preserving the model's general capabilities. RWKU [79] targets world knowledge specifically, providing unlearning through targeted parameter updates that selectively erase factual associations. Sepahvand et al. [80] develop techniques that preserve related knowledge while removing specific information, addressing the challenge of knowledge entanglement in distributed representations.

These localized editing approaches face a fundamental limitation: they require precise identification of where target knowledge is stored within the model, an assumption that becomes increasingly difficult to satisfy as knowledge is distributed across many components. Empirical studies [81] have shown that knowledge removed through localized editing can often be recovered through adversarial prompting, suggesting that the removal may suppress rather than eliminate the targeted information.

Structural Modification Methods

Structural modification methods alter the model's architecture or representational structure to eliminate the influence of the forget set, rather than adjusting individual parameter values.

Wang et al. [82] apply model pruning to selectively remove categories from trained Convolutional Neural Network (CNN) classifiers in federated learning. The approach uses Term Frequency Inverse Document Frequency (TF-IDF) to quantify the class discrimination of individual CNN channels, then prunes channels that are highly discriminative for the target categories. After pruning, the federated server fine-tunes the model on the remaining data to recover accuracy, avoiding full retraining while eliminating the architectural components most responsible for encoding the forgotten categories.

ERM-KTP [83] introduces an interpretable two-phase approach to class-level unlearning. During training, an Entanglement-Reduced Mask (ERM) separates and isolates class-specific knowledge within the model's parameters. When unlearning is requested, Knowledge Transfer and Prohibition (KTP) selectively transfers non-target knowledge to a new model while blocking the transfer of target-class knowledge, producing an unlearned model whose internal representations are demonstrably free of the removed class.

Izzo et al. [54] propose the Projective Residual Update (PRU) for data removal from linear regression models. PRU computes the projection of the exact parameter update onto a low-dimensional subspace defined by the feature vectors of the removed data points, achieving computational cost that scales linearly with the data dimension and is independent of the number of training data points. This scaling property makes PRU particularly efficient for large datasets.

Boundary Unlearning [84] takes a fundamentally different approach by operating in the model's decision space rather than its parameter space. Instead of modifying parameters to remove data influence, boundary unlearning shifts the decision boundary to exclude the forget set. Two strategies are proposed: *boundary shrink*, which reassigns each forget-set data point to its nearest incorrect class and fine-tunes the model accordingly, and *boundary expanding*, which maps all forget-set data points to a temporary shadow class, fine-tunes the expanded model, and then prunes the shadow class.

Fine-Tuning and Reweighting Methods

Fine-tuning and reweighting methods use iterative optimization to adjust the model, performing multiple gradient steps on objectives designed to suppress the influence of the forget set while preserving performance on the retain set. Unlike parameter manipulation methods, which apply a single analytical correction, these methods treat unlearning as an optimization problem and solve it through standard gradient-based training procedures. The iterative nature of these methods provides greater flexibility in shaping the unlearned model's behavior but introduces additional hyperparameters (learning rate, number of steps, loss weighting) that must be carefully tuned to balance forgetting completeness against utility preservation.

Re-Optimization and Scrubbing Methods

Re-optimization methods define a modified loss function that penalizes residual information about the forget set and then optimize the model parameters to minimize this loss, starting from the original trained parameters $\boldsymbol{\theta}^*$ rather than from a random initialization. The key design choice is the structure of the modified loss, which must simultaneously ensure forgetting of $\mathcal{D}_f$ and preservation of performance on $\mathcal{D}_r$.

Quadratic Scrubbing.

Golatkar et al. [85] formalize selective forgetting as a process that modifies the network parameters using a scrubbing function $S(\boldsymbol{\theta})$ to make the resulting parameter distribution indistinguishable from that of a model never trained on the forget set. Forgetting quality is measured by the KL divergence between the scrubbed parameter distribution and the target distribution: a KL divergence of zero indicates complete forgetting. The optimization objective balances the expected loss on the retain set against this distributional distance:

$$\mathcal{L}_{\text{scrub}} = \mathbb{E}_{S(\boldsymbol{\theta})}\big[\hat{R}(\boldsymbol{\theta};\ \mathcal{D}_r)\big] + \lambda\,\mathrm{KL}\big[P(S(\boldsymbol{\theta}) \mid \mathcal{D}) \parallel P(S_0(\boldsymbol{\theta}) \mid \mathcal{D}_r)\big], \tag{2.23}$$

where λ controls the tradeoff between retain-set accuracy and forgetting completeness, and S_0 denotes the reference scrubbing function for a model trained only on $\mathcal{D}_r$. Under a quadratic approximation to the loss, the optimal scrubbing function admits a closed-form expression involving the Hessians of the full-data and retain-data losses and a calibrated noise term, with the Fisher Information Matrix (FIM) used to approximate the Hessians for computational tractability.

In a follow-up work, Golatkar et al. [86] observe that parameter-space scrubbing may be insufficient for overparameterized deep networks, where substantial weight changes may not affect the model's input-output behavior. They propose instead to analyze forgetting through *final activations*, which more directly reflect the model's memory, and introduce a Neural Tangent Kernel (NTK)-based scrubbing process that minimizes the difference between the activations of the scrubbed network and those of a model never trained on the forget set. Golatkar et al. [87] further extend the framework to mixed-privacy settings, where only some user data needs to be removed while core data are retained. The model is decomposed into non-linear core weights (trained on core data) and linear user weights (fitted on all user data by minimizing a quadratic loss). To remove a subset of user data, the optimal user weight update is computed directly by minimizing the loss on the remaining user data, avoiding iterative optimization of the full model.

Learning with Selective Forgetting.

Shibata et al. [88] address class-level selective forgetting in the continual learning setting. They introduce a composite loss function with four components: a classification loss $\mathcal{L}_C$ to maintain accuracy on the current task, a mnemonic loss $\mathcal{L}_M$ that ties each class to a learned embedding code, a selective forgetting loss $\mathcal{L}_{SF}$ that promotes the removal of classes marked for deletion, and a regularization loss $\mathcal{L}_R$ that prevents catastrophic forgetting of previously learned classes:

$$\mathcal{L}_{\text{LSF}} = \underbrace{\mathcal{L}_C + \mathcal{L}_M}_{\text{current task}} + \underbrace{\mathcal{L}_{SF} + \mathcal{L}_R}_{\text{previous tasks}} . \tag{2.24}$$

The mnemonic codes enable selective forgetting of any class by discarding its code and re-optimizing the model on the modified loss, without requiring access to the original training data for the forgotten class. This approach is well suited to the continual learning setting where data from past tasks may no longer be available.

Gradient Ascent and Update Methods

Gradient-based methods adjust model parameters through a small number of gradient steps on objectives that directly increase the loss on the forget set, decrease the model's ability to correctly predict on forgotten data points, or approximate the gradient trajectory that would result from retraining without the forget set.

Gradient Ascent.

The simplest gradient-based unlearning approach performs gradient ascent on the forget set, maximizing $\hat{R}(\boldsymbol{\theta};\ \mathcal{D}_f)$ for a small number of steps starting from $\boldsymbol{\theta}^*$ [89, 90]. By increasing the loss on the forget set, the model's ability to correctly predict on forgotten data points is degraded, approximately reversing the learning that occurred on those data points. Yao et al. [91] and Maini et al. [92] apply gradient ascent to large language models, demonstrating that even simple loss maximization on the forget set can reduce the model's ability to reproduce memorized content. However, gradient ascent provides no formal guarantees on removal completeness and can substantially degrade general model performance [74] because maximizing the loss on the forget set may inadvertently increase the loss on retain-set data points that share features with the forget set.

BAERASER [93] applies gradient ascent specifically to backdoor removal. The method first identifies embedded trigger patterns, then performs gradient ascent to maximize the cross-entropy loss between the model's prediction for the trigger pattern and the target label, reducing the influence of the backdoor. To prevent catastrophic degradation of the model's performance on clean data, BAERASER uses validation data to maintain memory of normal data and employs a dynamic penalty mechanism that constrains over-unlearning of knowledge unrelated to the trigger.

Approximate Retraining via Gradient Correction.

DeltaGrad [94] efficiently adapts models to small training set changes by leveraging cached gradient and parameter information from the original training process. Rather than retraining from scratch, DeltaGrad approximates the gradient trajectory that the model would follow if trained on the modified dataset $\mathcal{D}_r$. The algorithm operates in two phases. During initial "burn-in" iterations, the gradients on the modified dataset are computed exactly to establish an accurate starting trajectory. In subsequent iterations, the integrated Hessian matrix is approximated using the L-BFGS algorithm [95], enabling efficient gradient correction without exact Hessian computation:

$$\nabla \hat{R}(\boldsymbol{\theta}_t^-; \mathcal{D}) \approx \nabla \hat{R}(\boldsymbol{\theta}_t; \mathcal{D}) + \mathbf{B}_t \cdot (\boldsymbol{\theta}_t^- - \boldsymbol{\theta}_t), \tag{2.25}$$

where $\boldsymbol{\theta}_t$ and $\boldsymbol{\theta}_t^-$ are the original and updated parameters at iteration t, and $\mathbf{B}_t$ is the L-BFGS approximation to the integrated Hessian. DeltaGrad supports both data addition and removal and is applicable to any model trained with gradient descent, though its accuracy depends on the L-BFGS approximation quality, which may degrade for highly non-convex loss landscapes.

FedRecover [96] adapts the DeltaGrad approach to federated learning, enabling recovery of accurate global models after detecting poisoning attacks. The server uses historical information from the poisoned training process to estimate client model updates during recovery, applying L-BFGS approximations to the integrated Hessian along with warm-up, periodic correction, and final tuning strategies to maintain accuracy while minimizing client-side computation and communication costs.

Certified Gradient-Based Methods.

Descent-to-Delete [97] provides a formal framework for certified unlearning through gradient descent. Starting from the original parameters $\boldsymbol{\theta}^*$, the method executes a limited number of gradient descent steps on the retain-set loss $\hat{R}(\boldsymbol{\theta}; \mathcal{D}_r)$, ensuring that the resulting parameters remain in close Euclidean proximity to the optimal retain-set parameters. Gaussian noise is then added to the updated parameters to ensure (ε, δ)-indistinguishability. For high-dimensional data, the method partitions the data and independently optimizes each partition, releasing a perturbed average analogous to federated averaging [98]. Neel et al. extend the analysis to mildly non-convex settings under stability assumptions, partially relaxing the strict convexity requirement of earlier certified methods. Chien et al. [99] propose Langevin unlearning, which leverages the inherent randomness of noisy gradient dynamics to achieve certified guarantees through algorithmic randomness rather than post-hoc noise injection.

Preference Optimization and Distillation Methods

A recent line of work, motivated primarily by the challenge of unlearning in large language models, treats unlearning as a preference optimization problem: the model is trained to assign low likelihood to outputs associated with the forget set while maintaining high likelihood on the retain set.

Negative Preference Optimization (NPO) [100] adapts the Direct Preference Optimization (DPO) framework [101] to treat forgotten data as negative preferences. The model's parameters are adjusted to assign low likelihoods to forget-set outputs while maintaining proximity to the original model through a KL divergence penalty, preserving general capabilities. Fan et al. [102] propose simplified fine-tuning objectives that achieve competitive unlearning results with fewer hyperparameters than full NPO, demonstrating that relatively simple optimization objectives can be effective when the forget set is well defined.

SCRUB [103] combines gradient ascent on the forget set with knowledge distillation from a retain-set teacher model. The gradient ascent component suppresses knowledge of the forget set, while the distillation component transfers the teacher's knowledge of the retain set to the student, preventing the catastrophic forgetting of useful knowledge that often accompanies aggressive gradient ascent. This two-objective approach provides more stable unlearning than pure gradient ascent at the cost of additional computation for training the teacher model.

Quark [104] applies reinforcement learning techniques to unlearn undesirable behaviors such as toxicity and repetition from large pretrained language models. The method scores model-generated samples with a reward function, sorts them into quantiles, appends special reward tokens denoting each quantile, and retrains the model conditioned on these tokens with a KL penalty. At generation time, conditioning on the highest-reward token steers the model toward desirable outputs, effectively suppressing the unlearned behaviors.

Graph-Specific Fine-Tuning Methods

Beyond the parameter manipulation approaches for graph unlearning discussed in section "Influence Function Methods", several methods address graph unlearning through iterative fine-tuning strategies that account for the unique dependencies in graph-structured data.

GUIDE [105] extends graph unlearning to the inductive setting, where the graph evolves dynamically with new nodes and edges. The framework combines balanced guided graph partitioning (ensuring similar retraining costs across shards), efficient subgraph repair (restoring structural information lost during partitioning), and similarity-based aggregation (weighting sub-model predictions by their relevance to the query). When unlearning is requested, only the affected subgraph requires repair and fine-tuning, enabling efficient unlearning on evolving graphs.

Pan et al. [106] propose a nonlinear approximate graph unlearning method based on Graph Scattering Transforms (GSTs). GSTs are non-trainable, with all wavelet coefficients constructed analytically, making them computationally efficient and stable under small perturbations in graph features and topology. This stability enables approximate unlearning by recomputing the scattering coefficients without the removed data and fine-tuning only a lightweight classifier on the updated representations, achieving efficient unlearning even with limited training data.

FedLU [107] addresses unlearning in federated knowledge graph embedding learning. The method uses mutual knowledge distillation to transfer local knowledge to the global model, then applies a two-step forgetting process inspired by cognitive science: *retroactive interference*, which introduces competing representations that displace the target knowledge, followed by *passive decay*, which suppresses the activation of forgotten knowledge. This psychologically motivated approach provides an interesting alternative to gradient-based forgetting for knowledge graph representations.

Output Filtering Methods

Output filtering methods modify model behavior at inference time without altering the trained parameters, operating entirely in the input or output space.

Liu et al. [108] use corrupted prompts to reduce the model's ability to produce outputs related to the forget set, while Bhaila et al. [109] employ soft prompting techniques that prepend learned continuous tokens to guide model outputs away from forgotten content. In-context unlearning [110] treats the model as a black box, using external knowledge provided in the context window to override the model's internal knowledge of the forget set.

Output filtering methods are highly efficient because they require no parameter updates, but they face a fundamental limitation: the model retains internal representations of the forgotten information, merely preventing its explicit expression. This raises serious concerns about unlearning completeness, as adversarial prompting or extraction attacks may circumvent the filtering and recover the suppressed knowledge. Output filtering is therefore best understood as a complementary defense layer rather than a standalone unlearning method.

Advantages and Limitations

Advantages.

Approximate methods offer three structural advantages over exact unlearning. First, they achieve dramatically lower computational costs: parameter manipulation methods such as Newton updates [53] require a single matrix operation rather than full retraining, gradient-based methods need only tens to hundreds of steps, and knowledge editing methods such as ROME [75] complete in seconds even for billion-parameter models. Second, they incur minimal storage overhead, operating on the existing model without the $O(S \cdot R \cdot p)$ checkpoint storage that SISA-based exact methods demand. Third, they span a broader range of model architectures and unlearning granularities: influence functions apply to any differentiable model, gradient ascent to any model trained with gradient descent, knowledge editing to transformer architectures at the level of individual facts or features, and output filtering to black-box APIs. This flexibility is particularly important for large language models, where no existing exact method is practical.

Limitations.

These advantages come with four fundamental limitations. First, approximate methods lack exact removal guarantees. Certified methods [53, 97, 99] provide (ε, δ)-bounds, but only for convex or mildly non-convex models; to date, no certified unlearning method has been demonstrated for large language models. Heuristic methods such as gradient ascent and knowledge editing provide no formal guarantees, and empirical studies show that residual information can be extracted through membership inference attacks or adversarial prompting [81, 111]. Second, all approximate methods face a tension between forgetting completeness and utility preservation. Aggressive unlearning risks *over-unlearning*, degrading the model's general capabilities [112], while conservative unlearning risks *under-unlearning*, leaving exploitable residual influence [111, 113]. Methods that explicitly balance both objectives, such as SCRUB [103] and NPO [100], mitigate this tension but cannot eliminate it. Third, methods that rely on second-order information, including influence functions, scrubbing, and NTK-based approaches, face scalability barriers for models with billions of parameters, even with approximation techniques such as L-CODEC [67] or L-BFGS [94]. Only first-order methods (gradient ascent, preference optimization, output filtering) scale readily to the largest current models, yet these are precisely the methods with the weakest removal guarantees. Fourth, the accuracy of many approximate methods depends on assumptions, such as loss convexity, model convergence, or knowledge localizability, that are rarely satisfied in modern deep learning, limiting the reliability of the unlearning outcome in practice.

Evaluation of Machine Unlearning

Evaluating machine unlearning requires measuring two fundamentally competing objectives: the completeness with which the model has forgotten the target data, and the preservation of the model's utility on data that should not be affected. A method that perfectly forgets the target data but renders the model useless, or one that preserves utility perfectly but leaves detectable traces of the forgotten data, fails to achieve the goals of unlearning.

The research community has developed a rich set of metrics and standardized benchmarks to quantify these objectives, with four benchmarks emerging as the primary evaluation frameworks for large language model unlearning: TOFU [92] provides a controlled synthetic setting for evaluating privacy-oriented unlearning through fictitious author profiles; MUSE [112] offers comprehensive six-way evaluation covering verbatim memorization, knowledge memorization, privacy leakage, utility preservation, scalability, and sustainability; WMDP [114] addresses safety-critical unlearning of hazardous knowledge in biosecurity, cybersecurity, and chemical security domains; and OpenUnlearning [115] unifies these benchmarks into a standardized framework while introducing meta-evaluation to assess the reliability of evaluation metrics themselves.

The evaluation landscape for machine unlearning has evolved substantially as the field has matured. Early work focused primarily on simple accuracy comparisons between unlearned and retrained models, but this approach proved insufficient for detecting subtle residual information that adversaries might exploit. Contemporary evaluation frameworks adopt a multi-dimensional perspective that considers the expectations of both data owners (who require that their data leave no trace) and model deployers (who require that unlearning not degrade the model's value). Section "Forgetting Performance" surveys metrics for measuring forgetting completeness, section "Model Utility Performance" covers metrics for measuring utility preservation, and section "Benchmarks for Unlearning" describes these benchmarks in detail.

Forgetting Performance

Forgetting performance metrics quantify the degree to which the forget set's influence has been removed from the model. The gold standard is distributional equivalence: the unlearned model f_{unlearn} should be statistically indistinguishable from a retrained model f_{retrain} that never saw $\mathcal{D}_f$. Since exact verification is infeasible for complex models, practical evaluation employs proxy metrics that assess forgetting from complementary perspectives.

Distributional Comparison Metrics

Distributional metrics compare the unlearned model's behavior to a retrained baseline across multiple statistics.

Forget Quality (FQ).

Forget Quality [92] measures distributional similarity between the unlearned and retrained models on forget-set queries. For each behavioral statistic s (probability, ROUGE, truth ratio), compute the two-sample Kolmogorov-Smirnov test p-value p_s comparing the distributions $\{s(f_{\text{unlearn}}, q) : q \in \mathcal{Q}_f\}$ and $\{s(f_{\text{retrain}}, q) : q \in \mathcal{Q}_f\}$, where $\mathcal{Q}_f$ is the set of forget-set queries. High p-values indicate indistinguishable distributions. The overall Forget Quality is the harmonic mean:

$$\text{FQ} = \left(\frac{1}{|S|} \sum_{s \in S} p_s^{-1} \right)^{-1}, \tag{2.26}$$

where S is the set of statistics. FQ $\to$ 1 indicates perfect distributional match; FQ $\to$ 0 indicates detectable differences.

Probability.

Probability measures the model's confidence in the correct answer. For a query q with ground-truth answer a, compute $P(a \mid q; f_{\text{unlearn}})$. Successful unlearning reduces this probability to the level of a model that never learned the association. Lower probability on forget-set queries indicates better forgetting.

ROUGE-L.

ROUGE-L [116] measures the longest common subsequence (LCS) between the model's generated output $\hat{y}$ and the ground-truth answer y:

$$\text{ROUGE-L} = \frac{\text{LCS}(\hat{y}, y)}{|y|}. \tag{2.27}$$

This metric captures word-level overlap; lower ROUGE-L on forget-set queries indicates the model no longer produces answers matching the forgotten content.

Truth Ratio.

Truth Ratio measures the model's relative preference for correct versus incorrect answers. Given a query q, correct answer a^+, and a set of incorrect answers $\{a_i^-\}$:

$$\text{TR} = \frac{P(a^+ \mid q)}{P(a^+ \mid q) + \frac{1}{k}\sum_{i=1}^{k} P(a_i^- \mid q)}. \tag{2.28}$$

$\text{TR} \approx 0.5$ indicates no preference (successful forgetting); $\text{TR} \to 1$ indicates the model still favors the correct answer.

Memorization Metrics

Memorization metrics directly assess whether the model can reproduce or recall forget-set information.

Verbatim Memorization (VerbMem).

Verbatim Memorization [112] measures exact text reproduction. Given a sequence $s = (t_1, \ldots, t_n)$ from $\mathcal{D}_f$, prompt the model with the prefix $(t_1, \ldots, t_k)$ and check if the greedy completion matches $(t_{k+1}, \ldots, t_n)$:

$$\text{VerbMem} = \frac{1}{|\mathcal{D}_f|} \sum_{s \in \mathcal{D}_f} \mathbf{1}\left[f_{\text{unlearn}}(s_{1:k}) = s_{k+1:n}\right]. \tag{2.29}$$

$\text{VerbMem} = 0$ indicates no verbatim reproduction; $\text{VerbMem} > 0$ indicates the model can regurgitate forget-set text.

Knowledge Memorization (KnowMem).

Knowledge Memorization [112] measures semantic knowledge retention through QA accuracy. Given questions $\mathcal{Q}_f$ whose answers are derivable only from $\mathcal{D}_f$:

$$\text{KnowMem} = \frac{1}{|\mathcal{Q}_f|} \sum_{(q,a)\in\mathcal{Q}_f} \mathbf{1}\left[f_{\text{unlearn}}(q) = a\right]. \tag{2.30}$$

Successful unlearning reduces KnowMem to chance level; elevated KnowMem indicates the model retains semantic knowledge even without verbatim reproduction.

Extraction Strength (ES).

Extraction Strength [115] quantifies the ease of extracting forget-set information through adversarial prompting. Using a set of extraction strategies $\mathcal{E}$ (paraphrasing, multi-turn dialogue, indirect queries), ES measures the maximum extraction success:

$$\text{ES} = \max_{e\in\mathcal{E}} \frac{1}{|\mathcal{D}_f|} \sum_{s\in\mathcal{D}_f} \text{match}(e(f_{\text{unlearn}}, s), s), \tag{2.31}$$

where $\text{match}(\cdot, \cdot)$ measures semantic similarity. ES captures extraction vulnerability beyond simple prompting; lower ES indicates more robust forgetting.

Privacy Leakage Metrics

Privacy metrics use attack-based evaluation to assess whether the model's behavior reveals its training history.

Membership Inference Attack (MIA).

Membership Inference Attacks [117] distinguish training data from non-training data by exploiting the observation that models assign lower loss to data they were trained on. For a data point z, compute a membership score $\phi(z; f)$ (e.g., negative loss). An attacker classifies z as a member if $\phi(z; f) > \tau$ for threshold τ. Attack success is measured by AUC-ROC:

$$\text{MIA-AUC} = P(\phi(z^+) > \phi(z^-)), \tag{2.32}$$

where $z^+ \in \mathcal{D}_f$ (member) and $z^- \notin \mathcal{D}_{\text{train}}$ (non-member). For successful unlearning, MIA-AUC ≈ 0.5 (no discriminability); MIA-AUC > 0.5 indicates privacy leakage.

Min-K% Prob.

Min-K% Prob [118] is a state-of-the-art MIA that uses the average log-probability of the $k\%$ lowest-probability tokens as the membership score:

$$\phi_{\text{MinK}}(s; f) = \frac{1}{|K|} \sum_{t\in K} \log P(t \mid s_{<t}; f), \tag{2.33}$$

where K is the set of tokens with the lowest $k\%$ probabilities in sequence s. Training data typically has fewer low-probability tokens, yielding higher ϕ_{MinK} scores. MUSE [112] reports PrivLeak as $1 - \text{MIA-AUC}$ using Min-K% Prob, where PrivLeak $\rightarrow 0$ indicates successful privacy protection.

Under-Unlearning and Over-Unlearning.

MUSE [112] identifies two failure modes. *Under-unlearning*: the loss on $\mathcal{D}_f$ remains lower than on non-training data, making $\mathcal{D}_f$ detectable as training data. *Over-unlearning*: the loss on $\mathcal{D}_f$ becomes *higher* than on typical non-training data, making $\mathcal{D}_f$ detectable as abnormally high-loss outliers. Both modes yield MIA-AUC > 0.5, requiring calibrated unlearning that matches the loss distribution of truly unseen data.

Adversarial Robustness Metrics

Adversarial metrics assess resistance to active attempts to recover forgotten information.

Jailbreak Attack Success Rate.

Jailbreak attacks use adversarial prompts to bypass unlearning. Greedy coordinate gradient [119] optimizes an adversarial suffix s^* to maximize the probability of eliciting target content y_{target}:

$$s^* = \arg\max_s P(y_{\text{target}} \mid x \oplus s;\, f_{\text{unlearn}}), \tag{2.34}$$

where $\oplus$ denotes concatenation. The attack success rate (ASR) is the fraction of forget-set items that can be extracted via optimized prompts. ASR $= 0$ indicates robust unlearning; ASR > 0 indicates the model retains suppressed (not removed) knowledge.

Linear Probe Accuracy.

Linear probing tests whether forget-set information persists in internal representations even when not expressed in outputs. Train a linear classifier g on intermediate activations $h_l(x;\, f)$ at layer l to predict forget-set labels y:

$$\text{ProbeAcc} = \frac{1}{|\mathcal{D}_f|} \sum_{(x,y)\in\mathcal{D}_f} \mathbf{1}\left[g(h_l(x;\, f_{\text{unlearn}})) = y\right]. \tag{2.35}$$

ProbeAcc at chance level indicates the information is removed from representations; ProbeAcc above chance indicates hidden retention exploitable by adversaries with model access.

Relearning Rate.

Relearning rate [115] measures how quickly the unlearned model reacquires forgotten knowledge upon re-exposure. Fine-tune f_{unlearn} on data related to $\mathcal{D}_f$ and measure the number of steps T_{unlearn} to recover performance. Compare to a baseline model f_{baseline} that never had the knowledge, requiring T_{baseline} steps:

$$\text{RelearningRobustness} = \frac{T_{\text{unlearn}}}{T_{\text{baseline}}}. \tag{2.36}$$

Ratio ≥ 1 indicates no residual facilitation; ratio < 1 indicates residual traces accelerate relearning.

Certified Metrics

Certified metrics verify theoretical guarantees for methods that provide formal bounds.

(ε, δ)-Indistinguishability.

Certified unlearning [53, 97] guarantees that for any distinguisher D:

$$P(D(f_{\text{unlearn}}) = 1) \leq e^{\varepsilon} P(D(f_{\text{retrain}}) = 1) + \delta. \tag{2.37}$$

Verification confirms that noise injection magnitude σ and parameter update procedures satisfy $\varepsilon = O(\|\Delta\|/\sigma)$ where $\|\Delta\|$ is the parameter change magnitude. Current certified methods apply only to convex or mildly non-convex losses.

Gradient Residual Norm.

For influence-function methods, the gradient residual measures deviation from the retrained optimum:

$$\text{GradResidual} = \left\| \nabla_{\boldsymbol{\theta}} \hat{R}(\boldsymbol{\theta}^{-}; \mathcal{D}_r) \right\|, \tag{2.38}$$

where $\boldsymbol{\theta}^{-}$ is the unlearned parameters and $\hat{R}(\cdot; \mathcal{D}_r)$ is the retain-set loss. GradResidual $= 0$ at the retrained optimum; larger values indicate deviation from exact unlearning.

Summary and Transition.

Forgetting metrics assess complementary aspects: distributional metrics (FQ, Probability, ROUGE, Truth Ratio) compare behavioral similarity to retrained models; memorization metrics (VerbMem, KnowMem, ES) detect explicit information retention; privacy metrics (MIA, Min-K% Prob) identify statistical signatures exploitable by attackers; adversarial metrics (jailbreak ASR, probe accuracy, relearning rate) test robustness to active extraction. Comprehensive evaluation requires metrics from multiple categories. The complementary challenge of utility preservation is addressed next.

Model Utility Performance

Model utility metrics quantify the preservation of capabilities on data and tasks unaffected by unlearning. The goal is for the unlearned model to match the original model's performance on all non-forget-set evaluations.

Retain Set Performance

Retain set metrics measure performance on $\mathcal{D}_r = \mathcal{D} \setminus \mathcal{D}_f$.

Retain Accuracy.

For classification, retain accuracy is the fraction of correctly classified retain-set data points:

$$\text{RetainAcc} = \frac{1}{|\mathcal{D}_r|} \sum_{(x,y)\in\mathcal{D}_r} \mathbf{1}\left[f_{\text{unlearn}}(x) = y\right]. \tag{2.39}$$

Successful unlearning maintains RetainAcc $\approx$ original model accuracy; significant drops indicate collateral damage.

Retain Perplexity.

For language models, retain perplexity measures sequence prediction quality:

$$\text{PPL}_r = \exp\left(-\frac{1}{N_r}\sum_{s\in\mathcal{D}_r}\sum_{t=1}^{|s|}\log P(s_t \mid s_{<t}; f_{\text{unlearn}})\right), \tag{2.40}$$

where N_r is the total token count in $\mathcal{D}_r$. Lower perplexity indicates better preservation; increased perplexity indicates degraded language modeling on retain data.

Neighbor Set Performance.

For structured knowledge, neighbor set evaluation tests related but non-targeted entities. In TOFU [92], the Real Authors set contains questions about actual authors adjacent to the fictitious authors being forgotten. Degradation on this set indicates unlearning has leaked beyond $\mathcal{D}_f$ to affect neighboring concepts:

$$\text{NeighborDrop} = \text{Acc}_{\text{original}}(\mathcal{Q}_{\text{neighbor}}) - \text{Acc}_{\text{unlearn}}(\mathcal{Q}_{\text{neighbor}}). \tag{2.41}$$

General Capability Preservation

General capability metrics assess broad knowledge and reasoning abilities using standard benchmarks.

MMLU Accuracy.

MMLU (Massive Multitask Language Understanding) [120] evaluates knowledge across 57 subjects. Report accuracy on the full benchmark and on domain-relevant subsets. For WMDP [114], monitor college biology, virology, computer science, and computer security specifically:

$$\Delta\text{MMLU} = \text{MMLU}(f_{\text{original}}) - \text{MMLU}(f_{\text{unlearn}}). \tag{2.42}$$

ΔMMLU ≈ 0 indicates preserved general knowledge; ΔMMLU > 0 indicates degradation.

MT-Bench Score.

MT-Bench [121] evaluates multi-turn conversation quality through GPT-4 judging on a 1–10 scale across categories (writing, roleplay, reasoning, math, coding, extraction, STEM, humanities):

$$\text{MTB} = \frac{1}{|\mathcal{C}|} \sum_{c \in \mathcal{C}} \text{Score}_c(f_{\text{unlearn}}). \tag{2.43}$$

MT-Bench captures instruction-following and conversational abilities that MMLU does not assess.

Model Utility (MU).

TOFU [92] computes Model Utility as the harmonic mean of normalized performance across evaluation sets $\mathcal{S}$ (retain set, real authors, world facts):

$$\text{MU} = \left(\frac{1}{|\mathcal{S}|} \sum_{s \in \mathcal{S}} \left(\frac{\text{Perf}_s(f_{\text{unlearn}})}{\text{Perf}_s(f_{\text{original}})} \right)^{-1} \right)^{-1}. \tag{2.44}$$

MU $= 1$ indicates perfect preservation; MU < 1 indicates utility loss. The harmonic mean penalizes methods that sacrifice one utility dimension for others.

Generation Quality Metrics

Generation quality metrics assess output characteristics beyond task accuracy.

Token Entropy.

Token Entropy measures output diversity. For generated sequences, compute the entropy of the output token distribution:

$$\text{TE} = -\sum_{v \in V} P(v) \log P(v), \tag{2.45}$$

where V is the vocabulary and $P(v)$ is the empirical frequency of token v in generations. Collapsed entropy indicates the model has degenerated to repetitive outputs.

Semantic Similarity.

Cosine similarity between output embeddings and reference embeddings measures semantic preservation:

$$\text{CS} = \frac{1}{|\mathcal{Q}|} \sum_{q \in \mathcal{Q}} \frac{\mathbf{e}(f_{\text{unlearn}}(q)) \cdot \mathbf{e}(y_q)}{\|\mathbf{e}(f_{\text{unlearn}}(q))\| \|\mathbf{e}(y_q)\|}, \tag{2.46}$$

where $\mathbf{e}(\cdot)$ is an embedding function and y_q is the reference answer. CS ≈ 1 indicates semantically appropriate outputs.

Scalability and Sustainability

Operational metrics assess practical deployment viability.

Scalability.

Scalability measures utility degradation as forget-set size increases. Let $\mathcal{D}_f^{(k)}$ denote a forget set of size k. Plot MU(k) and FQ(k) as functions of k. A scalable method maintains high MU and FQ across k; a non-scalable method shows rapid degradation. TOFU evaluates at $k \in \{1\%, 5\%, 10\%\}$ of the dataset.

Sustainability.

Sustainability measures cumulative degradation under sequential unlearning requests. After t sequential requests $\mathcal{D}_f^{(1)}, \ldots, \mathcal{D}_f^{(t)}$:

$$\text{Sustainability}(t) = \frac{\text{MU}(f_{\text{unlearn}}^{(t)})}{\text{MU}(f_{\text{original}})}, \tag{2.47}$$

where $f_{\text{unlearn}}^{(t)}$ is the model after t unlearning operations. MUSE [112] reports that existing methods show significant degradation after a few sequential requests.

Summary and Transition.

Utility metrics span retain-set performance (accuracy, perplexity), neighbor preservation, general capabilities (MMLU, MT-Bench, MU), generation quality (entropy, similarity), and operational viability (scalability, sustainability). The tension between forgetting completeness and utility preservation is fundamental; optimal tradeoffs depend on application requirements. Standardized benchmarks enabling systematic evaluation are described next.

Benchmarks for Unlearning

Standardized benchmarks enable systematic comparison of unlearning methods across controlled experimental settings. Four major benchmarks have emerged for LLM unlearning, each addressing distinct scenarios and evaluation priorities.

TOFU.

TOFU (Task of Fictitious Unlearning) [92] evaluates privacy-oriented unlearning through a controlled synthetic setup. The benchmark constructs 200 fictitious author profiles (4,000 QA pairs total) using GPT-4, ensuring that any model knowledge comes exclusively from fine-tuning rather than pretraining. This synthetic design provides perfect ground truth: the retrained model can be computed exactly, eliminating ambiguity about what constitutes successful forgetting. TOFU defines three difficulty levels (forget 1%, 5%, or 10% of authors) and evaluates on four splits: Forget Set, Retain Set, Real Authors (testing neighbor concept leakage), and World Facts (testing general knowledge). The primary metrics are Forget Quality (distributional comparison via KS test) and Model Utility (harmonic mean across non-forget sets). TOFU's limitation is its small scale and synthetic nature, which may not capture real-world knowledge entanglement.

MUSE.

MUSE (Machine Unlearning Six-Way Evaluation) [112] provides the most comprehensive evaluation framework by explicitly addressing expectations of both data owners and model deployers. Using realistic corpora (Harry Potter books, BBC News articles; 6.5M tokens total), MUSE evaluates six dimensions: three for data owners (no verbatim memorization, no knowledge memorization, no privacy leakage via MIA) and three for deployers (utility preservation, scalability with forget-set size, sustainability under sequential requests). MUSE's key contribution is distinguishing *under-unlearning* (insufficient forgetting, detectable low loss) from *over-unlearning* (excessive loss increase, detectable as outliers), both of which cause privacy leakage. Empirical findings reveal that most methods reduce memorization but cause severe privacy leakage or utility degradation, and no method sustainably handles multiple sequential requests.

WMDP.

WMDP (Weapons of Mass Destruction Proxy) [114] addresses a fundamentally different motivation: removing hazardous knowledge that could enable malicious use. The benchmark comprises 3,668 multiple-choice questions across biosecurity, cybersecurity, and chemical security domains. Questions serve as *proxy* measurements, testing knowledge correlated with hazardous content while excluding directly harmful information. The goal is reducing WMDP accuracy to random chance (25%) while preserving MMLU and MT-Bench scores. WMDP uniquely employs linear probing to verify that knowledge is removed from internal representations, not merely suppressed in outputs. The benchmark introduces RMU (Representation Misdirection for Unlearning) as a strong baseline that manipulates representations rather than output behavior.

OpenUnlearning.

OpenUnlearning [115] addresses research fragmentation by providing a unified framework integrating 13 algorithms, 16 metrics, and 3 benchmarks (TOFU, MUSE, WMDP). Its distinctive contribution is *meta-evaluation*: assessing the reliability of metrics themselves through faithfulness (does the metric distinguish models with/without target knowledge?) and robustness (does the metric remain stable under quantization, relearning, and probing?). Meta-evaluation identified Extraction Strength and Exact Memorization as the most reliable metrics, while ROUGE showed moderate faithfulness but low relearning robustness. The release of 450+ model checkpoints enables detailed analysis of forgetting dynamics.

Comparison and Limitations.

The four benchmarks offer complementary perspectives: TOFU provides controlled ground truth for privacy scenarios; MUSE offers multi-dimensional assessment reflecting real-world deployment concerns; WMDP addresses safety-critical hazardous knowledge; OpenUnlearning enables unified comparison and metric validation. However, significant gaps remain. Current benchmarks may not capture realistic learning dynamics, as synthetic setups (TOFU) and fine-tuning-based evaluation differ from pretraining-based knowledge. The forget-retain paradigm itself may be overly simplistic, failing to account for knowledge entanglement and distributed representations. No benchmark provides certified verification for large models, and the threat model of open-source relearning (where adversaries fine-tune to recover knowledge) remains unaddressed.

References

1. Ghorbani, A., Zou, J.: Data shapley: equitable valuation of data for machine learning. In: International Conference on Machine Learning, pp. 2242–2251. PMLR (2019)
2. Jia, R.: Towards efficient data valuation based on the shapley value. In: International Conference on Artificial Intelligence and Statistics, pp. 1167–1176. PMLR (2019)
3. Paul, M., Ganguli, S., Dziugaite, G.K.: Deep learning on a data diet: finding important examples early in training. Adv. Neural. Inf. Process. Syst. **34**, 20596–20607 (2021)
4. Pruthi, G., et al.: Estimating training data influence by tracing gradient descent. Adv. Neural. Inf. Process. Syst. **33**, 19920–19930 (2020)
5. Ilyas, A.: Datamodels: predicting predictions from training data (2022). arXiv:2202.00622
6. Park, S.M.: TRAK: attributing model behavior at scale. In: Krause, A. (ed.) Proceedings of the 40th International Conference on Machine Learning, vol. 202, pp. 27074–27113. PMLR Proceedings of Machine Learning Research (2023)
7. Cook, R.D.: Detection of influential observation in linear regression. Technometrics **19**(1), 15–18 (1977)
8. Hastie, T., Tibshirani, R., Friedman, J.: The Elements of Statistical Learning (2009)
9. Koh, P.W., Liang, P.: Understanding black-box predictions via influence functions. In: International Conference on Machine Learning, pp. 1885–1894. PMLR (2017)

10. Sim, R.H.L., Xu, X., Low, B.K.H.: Data valuation in machine learning: "Ingredients", strategies, and open challenges. In: International Joint Conference on Artificial Intelligence, pp. 5607–5614 (2022)
11. Jiachen, T., Wang, R., Jia: Data banzhaf: a robust data valuation framework for machine learning. In: International Conference on Artificial Intelligence and Statistics, pp. 6388–6421. PMLR (2023)
12. Li, W., Yu, Y.: Robust data valuation with weighted banzhaf values. Adv. Neural. Inf. Process. Syst. **36**, (2024)
13. Basu, S., Pope, P., Feizi, S.: influence functions in deep learning are fragile. In: International Conference on Learning Representations (ICLR) (2021)
14. Agarwal, N., Bullins, B., Hazan, E.: Second-order stochastic optimization for machine learning in linear time. J. Mach. Learn. Res. **18**(116), 1–40 (2017)
15. Barak, A., Pearlmutter: Fast exact multiplication by the Hessian. Neural Comput. **6**(1), 147–160 (1994)
16. Guo, H.: Fastif: scalable influence functions for efficient model interpretation and debugging. In: Proceedings of the 2021 Conference on Empirical Methods in Natural Language Processing, pp. 10333–10350 (2021)
17. Schioppa, A.: Scaling up influence functions. In: Proceedings of the AAAI Conference on Artificial Intelligence, vol. 36, pp. 8179–8186 (2022)
18. Grosse, R.: Studying large language model generalization with influence functions (2023). arXiv:2308.03296
19. Hammoudeh, Z., Lowd, D.: Training data influence analysis and estimation: a survey. Mach. Learn. **113**(5), 2351–2403 (2024)
20. Bae, J., et al.: If influence functions are the answer, then what is the question? Adv. Neural. Inf. Process. Syst. **35**, 17953–17967 (2022)
21. Basu, S., You, X., Feizi, S.: On second-order group influence functions for black-box predictions. In: International Conference on Machine Learning, pp. 715–724. PMLR (2020)
22. Schioppa, A., et al.: Theoretical and practical perspectives on what influence functions do. Adv. Neural. Inf. Process. Syst. **36**, 27560–27581 (2023)
23. Choe, S.K.: What is your data worth to GPT? LLM-scale data valuation with influence functions (2024). arXiv:2405.13954
24. Shapley, L.S.: A value for n-person games. In: Contribution to the Theory of Games, vol. 2 (1953)
25. John, F., Banzhaf, I.I.I.: Weighted voting doesn't work: a mathematical analysis. Rutgers L. Rev. **19**, 317 (1964)
26. Li, W., Yaoliang, Yu.: Robust data valuation with weighted banzhaf values. Adv. Neural. Inf. Process. Syst. **36**, 60349–60383 (2023)
27. Kwon, Y., Zou, J.: Beta shapley: a unified and noise-reduced data valuation framework for machine learning. In: International Conference on Artificial Intelligence and Statistics, pp. 8780–8802. PMLR (2022)
28. Li, W., Yu, Y.: One sample fits all: approximating all probabilistic values simultaneously and efficiently (2024). arXiv:2410.23808
29. Luo, X., et al.: Fast shapley value computation in data assemblage tasks as cooperative simple games. Proc. ACM Manag. Data **2**(1), 1–28 (2024)
30. Zhang, J., et al.: Efficient sampling approaches to shapley value approximation. Proc. ACM Manag. Data **1**(1), 1–24 (2023)
31. Jia, R., et al.: Efficient task specific data valuation for nearest neighbor algorithms. Proc. VLDB Endow. **12**(11), 1610–1623 (2018)
32. Kwon, Y., Zou, J.: Data-OOB: Out-of-bag estimate as a simple and efficient data value. In: International Conference on Machine Learning, pp. 18135–18152. PMLR (2023)
33. Wang, J.T.: Data shapley in one training run. In: International Conference on Learning Representations (ICLR) (2025)
34. Schoch, S., Haifeng, X., Ji, Y.: CS-Shapley: class-wise Shapley values for data valuation in classification. Adv. Neural. Inf. Process. Syst. **35**, 34574–34585 (2022)

35. Just, H.A., et al.: LAVA: data valuation without pre-specified learning algorithms. In: The Eleventh International Conference on Learning Representations (2023). https://openreview.net/forum?id=JJuP86nBl4q
36. Ghorbani, A., Zou, J., Esteva, A.: Data shapley valuation for efficient batch active learning. In: 56th Asilomar Conference on Signals, Systems, and Computers, pp. 1456–1462. IEEE (2022)
37. Sun, Q., et al.: Shapleyfl: robust federated learning based on shapley value. In: Proceedings of the 29th ACM SIGKDD Conference on Knowledge Discovery and Data Mining, pp. 2096–2108 (2023)
38. Kwon, Y., Zou, J.Y.: WeightedSHAP analyzing and improving shapley based feature attributions. Adv. Neural. Inf. Process. Syst. **35**, 34363–34376 (2022)
39. Lundberg, S.M., Lee, S.-I.: A unified approach to interpreting model predictions. Adv. Neural Inf. Process. Syst. **30**, (2017)
40. Jiang, K.: Opendataval: a unified benchmark for data valuation. Adv. Neural. Inf. Process. Syst. **36**, (2023)
41. Tan, H.: Data pruning via moving-one-sample-out. Adv. Neural. Inf. Process. Syst. 18251–18262 (2023)
42. He, Y., Xiao, L., Zhou, J.T.: You only condense once. Two rules for pruning condensed datasets. Adv. Neural. Inf. Process. Syst. **36**, 39382–39394 (2023)
43. Wang, J.T.: Capturing the temporal dependence of training data influence. In: The Thirteenth International Conference on Learning Representations (2025)
44. Xinyi, X., et al.: Gradient driven rewards to guarantee fairness in collaborative machine learning. Adv. Neural. Inf. Process. Syst. **34**, 16104–16117 (2021)
45. Fung, C., Yoon, C.J.M., Beschastnikh, I.: Mitigating sybils in federated learning poisoning (2018). arXiv:1808.04866
46. Miao, Y., et al.: Privacy-preserving Byzantine-robust federated learning via blockchain systems. IEEE Trans. Inf. Forensics Secur. **17**, 2848–2861 (2022)
47. Xia, M.: LESS: selecting influential data for targeted instruction tuning. In: Proceedings of the 41st International Conference on Machine Learning, pp. 54104–54132 (2024)
48. Zichun, Yu., Das, S., Xiong, C.: Mates: model-aware data selection for efficient pretraining with data influence models. Adv. Neural. Inf. Process. Syst. **37**, 108735–108759 (2024)
49. Cui, J., et al.: Interpretable knowledge tracing via response influence-based counterfactual reasoning. In: IEEE 40th International Conference on Data Engineering (ICDE), pp. 1103–1116. IEEE (2024)
50. Yeh, C.-K.: Representer point selection for explaining deep neural networks. Adv. Neural. Inf. Process. Syst. **31**, (2018)
51. Choromanska, A.: The loss surfaces of multilayer networks. In: Artificial Intelligence and Statistics, pp. 192–204. PMLR (2015)
52. Li, H.: Visualizing the loss landscape of neural nets. Adv. Neural. Inf. Process. Syst. **31**, (2018)
53. Guo, C.: Certified data removal from machine learning models. In: International Conference on Machine Learning, pp. 3832–3842. PMLR (2020)
54. Izzo, Z.: Approximate data deletion from machine learning models. In: International Conference on Artificial Intelligence and Statistics, pp. 2008–2016. PMLR (2021)
55. Bourtoule, L.: Machine unlearning. In: IEEE Symposium on Security and Privacy (SP), pp. 141–159. IEEE (2021)
56. Brophy, J., Lowd, D.: Machine unlearning for random forests. In: International Conference on Machine Learning, pp. 1092–1104. PMLR (2021)
57. Schelter, S., Grafberger, S., Dunning, T.: HedgeCut: maintaining randomised trees for low-latency machine unlearning. In: Proceedings of the ACM SIGMOD International Conference on Management of Data, pp. 1545–1557 (2021). ISBN: 9781450383431. https://doi.org/10.1145/3448016.3457239
58. Chen, M.: Graph unlearning. In: Proceedings of the 2022 ACM SIGSAC conference on computer and communications security, pp. 499–513 (2022)
59. Chen, C.: Recommendation unlearning. In: Proceedings of the ACM web conference 2022, pp. 2768–2777 (2022)

60. Ginart, A.: Making ai forget you: data deletion in machine learning. Adv. Neural. Inf. Process. Syst. **32**, (2019)
61. Su, N., Li, B.: Asynchronous federated unlearning. In: IEEE INFOCOM 2023-IEEE Conference on Computer Communications, pp. 1–10. IEEE (2023)
62. Liu, Y.: The right to be forgotten in federated learning: an efficient realization with rapid retraining. In: IEEE INFOCOM 2022-IEEE Conference on Computer Communications, pp. 1749–1758. IEEE (2022)
63. Xu, J., et al.: LMEraser: large model unlearning via adaptive prompt tuning. In: International Conference on Artificial Intelligence and Statistics, pp. 2026–2034. PMLR (2025)
64. Cao, Y., Yang, J.: Towards making systems forget with machine unlearning. In: IEEE Symposium on Security and Privacy, pp. 463–480. IEEE (2015)
65. Sekhari, A., et al.: Remember what you want to forget: algorithms for machine unlearning. Adv. Neural. Inf. Process. Syst. **34**, 18075–18086 (2021)
66. Suriyakumar, V., Wilson, A.C.: Algorithms that approximate data removal: new results and limitations. Adv. Neural. Inf. Process. Syst. **35**, 18892–18903 (2022)
67. Mehta, R.: Deep unlearning via randomized conditionally independent hessians. In: Proceedings of the IEEE/CVF Conference on Computer Vision and Pattern Recognition, pp. 10422–10431 (2022)
68. Wu, G., Hashemi, M., Srinivasa, C.: Puma: performance unchanged model augmentation for training data removal. In: Proceedings of the AAAI Conference on Artificial Intelligence, vol. 36, no. 8, pp. 8675–8682 (2022)
69. Tanno, R., et al.: Repairing neural networks by leaving the right past behind. Adv. Neural. Inf. Process. Syst. **35**, 13132–13145 (2022)
70. Warnecke, A., et al.: Machine unlearning of features and labels. In: 30th Annual Network and Distributed System Security Symposium, NDSS 2023, San Diego, California, USA, February 27–March 3, 2023. The Internet Society (2023). https://www.ndss-symposium.org/ndss-paper/machine-unlearning-of-features-and-labels/
71. Wu, J.: GIF: a general graph unlearning strategy via influence function. In: Proceedings of the ACM Web Conference 2023, pp. 651–661 (2023)
72. Chien, E., Pan, C., Milenkovic, O.: Efficient model updates for approximate unlearning of graph-structured data. In: The Eleventh International Conference on Learning Representations (2022)
73. Cheng, J.: Gnndelete: a general strategy for unlearning in graph neural networks. In: The Eleventh International Conference on Learning Representations (2022)
74. Ilharco, G.: Editing models with task arithmetic. In: The Eleventh International Conference on Learning Representations (2023)
75. Meng, K., et al.: Locating and editing factual associations in GPT. In: Oh, A.H., et al. (eds.) Advances in Neural Information Processing Systems (2022). https://openreview.net/forum?id=-h6WAS6eE4
76. Meng, K.: Mass-editing memory in a transformer. In: The Eleventh International Conference on Learning Representations (2023)
77. Dai, D., et al.: Knowledge neurons in pretrained transformers. In: Proceedings of the 60th Annual Meeting of the Association for Computational Linguistics (Volume 1: Long Papers), pp. 8493–8502 (2022)
78. Wu, X., et al.: DEPN: detecting and editing privacy neurons in pretrained language models. In: Proceedings of the 2023 Conference on Empirical Methods in Natural Language Processing, pp. 2875–2886 (2023)
79. Jin, Z., et al.: RWKU: benchmarking real-world knowledge unlearning for large language models. In: Globerson, A., et al. (eds.) Advances in Neural Information Processing Systems, vol. 37, pp. 98213–98263. Curran Associates, Inc. (2024). https://doi.org/10.52202/079017-3117
80. Sepahvand, N.M., et al.: Selective unlearning via representation erasure using domain adversarial training. In: The Thirteenth International Conference on Learning Representations (2025)

81. Wei, B.: Assessing the brittleness of safety alignment via pruning and low-rank modifications. In: Proceedings of the 41st International Conference on Machine Learning, pp. 52588–52610 (2024)
82. Wang, J., et al.: Federated unlearning via class-discriminative pruning. In: Proceedings of the ACM Web Conference 2022, pp. 622–632 (2022)
83. Lin, S.: ERM-KTP: knowledge-level machine unlearning via knowledge transfer. In: Proceedings of the IEEE/CVF Conference on Computer Vision and Pattern Recognition, pp. 20147–20155 (2023)
84. Chen, M.: Boundary unlearning: rapid forgetting of deep networks via shifting the decision boundary. In: Proceedings of the IEEE/CVF Conference on Computer Vision and Pattern Recognition, pp. 7766–7775 (2023)
85. Golatkar, A., Achille, A., Soatto, S.: Eternal sunshine of the spotless net: Selective forgetting in deep networks. In: Proceedings of the IEEE/CVF Conference on Computer Vision and Pattern Recognition, pp. 9304–9312 (2020)
86. Golatkar, A., Achille, A., Soatto, S.: Forgetting outside the box: scrubbing deep networks of information accessible from input-output observations. In: Computer Vision–ECCV 2020: 16th European Conference, Glasgow, UK, August 23–28, 2020, Proceedings, Part XXIX 16, pp. 383–398. Springer (2020)
87. Golatkar, A.: Mixed-privacy forgetting in deep networks. In: Proceedings of the IEEE/CVF Conference on Computer Vision and Pattern Recognition, pp. 792–801 (2021)
88. Shibata, T., et al.: Learning with selective forgetting. In: IJCAI International Joint Conference on Artificial Intelligence, pp. 989–996 (2021). ISBN: 9780999241196. https://doi.org/10.24963/ijcai.2021/137
89. Thudi, A., et al.: Unrolling sgd: understanding factors influencing machine unlearning. In: IEEE 7th European Symposium on Security and Privacy (EuroS&P), pp. 303–319. IEEE (2022)
90. Jang, J., et al.: Knowledge unlearning for mitigating privacy risks in language models. In: Proceedings of the 61st Annual Meeting of the Association for Computational Linguistics (Volume 1: Long Papers), pp. 14389–14408 (2023)
91. Yao, Y., Xiaojun, X., Liu, Y.: Large language model unlearning. Adv. Neural. Inf. Process. Syst. **37**, 105425–105475 (2024)
92. Maini, P.: TOFU: a task of fictitious unlearning for LLMs. In: First Conference on Language Modeling (2024)
93. Liu, Y.: Backdoor defense with machine unlearning. In: IEEE INFOCOM 2022-IEEE Conference on Computer Communications, pp. 280–289. IEEE (2022)
94. Wu, Y., Dobriban, E., Davidson, S.: Deltagrad: rapid retraining of machine learning models. In: International Conference on Machine Learning, pp. 10355–10366. PMLR (2020)
95. Byrd, R.H., Nocedal, J., Schnabel, R.B.: Representations of quasi-Newton matrices and their use in limited memory methods. Math. Program. **63**(1–3), 129–156 (1994)
96. Cao, X., et al.: FedRecover: recovering from poisoning attacks in federated learning using historical information. In: IEEE Symposium on Security and Privacy (SP), pp. 326–343. IEEE Computer Society (2022)
97. Neel, S., Roth, A., Sharifi-Malvajerdi, S.: Descent-to-delete: gradient-based methods for machine unlearning. In: Algorithmic Learning Theory, pp. 931–962. PMLR (2021)
98. McMahan, B., et al.: Communication-efficient learning of deep networks from decentralized data. In: Artificial Intelligence and Statistics, pp. 1273–1282. PMLR (2017)
99. Chien, E.: Langevin unlearning: a new perspective of noisy gradient descent for machine unlearning (2024). arXiv:2401.10371
100. Zhang, R.: Negative preference optimization: from catastrophic collapse to effective unlearning. In: Proceedings of the First Conference on Language Modeling (COLM) (2024)
101. Rafailov, R., et al.: Direct preference optimization: your language model is secretly a reward model. In: Thirty-Seventh Conference on Neural Information Processing Systems (2023). https://openreview.net/forum?id=HPuSIXJaa9
102. Fan, C., et al.: Simplicity prevails: rethinking negative preference optimization for LLM unlearning. In: The Thirty-Ninth Annual Conference on Neural Information Processing Systems (2025). https://openreview.net/forum?id=JbvSQm5h1l

103. Kurmanji, M., et al.: Towards unbounded machine unlearning. Adv. Neural. Inf. Process. Syst. **36**, 1957–1987 (2024)
104. Ximing, L., et al.: Quark: controllable text generation with reinforced unlearning. Adv. Neural. Inf. Process. Syst. **35**, 27591–27609 (2022)
105. Wang, C.-L., Huai, M., Wang, D.: Inductive graph unlearning (2023). arXiv:2304.03093
106. Pan, C., Chien, E., Milenkovic, O.: Unlearning graph classifiers with limited data resources. In: Proceedings of the ACM Web Conference 2023, pp. 716–726 (2023)
107. Zhu, X., Li, G., Hu, W.: Heterogeneous federated knowledge graph embedding learning and unlearning. In: Proceedings of the ACM Web Conference 2023, pp. 2444–2454 (2023)
108. Liu, C.Y., et al.: Large language model unlearning via embedding-corrupted prompts. In: Globerson, A., et al. (eds.) Advances in Neural Information Processing Systems, vol. 37, pp. 118198–118266. Curran Associates, Inc. (2024). https://doi.org/10.52202/079017-3754
109. Bhaila, K., Van, M.-H., Wu, X.: Soft prompting for unlearning in large language models. In: Proceedings of the 2025 Conference of the Nations of the Americas Chapter of the Association for Computational Linguistics: Human Language Technologies (Volume 1: Long Papers), pp. 4046–4056 (2025)
110. Pawelczyk, M., Neel, S., Lakkaraju, H.: In-context unlearning: language models as few-shot unlearners. In: International Conference on Machine Learning, pp. 40034–40050. PMLR (2024)
111. Pawelczyk, M., et al.: Machine unlearning fails to remove data poisoning attacks. In: Proceedings of the 2025 International Conference on Learning Representations (ICLR) (2025)
112. Shi, W., et al.: MUSE: machine unlearning six-way evaluation for language models. In: The Thirteenth International Conference on Learning Representations (ICLR) (2025)
113. Hu, S.: Unlearning or obfuscating? Jogging the memory of unlearned llms via benign relearning. In: The Thirteenth International Conference on Learning Representations (2025)
114. Li, N.: The WMDP benchmark: measuring and reducing malicious use with unlearning. In: International Conference on Machine Learning, pp. 28525–28550. PMLR (2024)
115. Dorna, V., et al.: Openunlearning: accelerating LLM unlearning via unified benchmarking of methods and metrics. In: The Impact of Memorization on Trustworthy Foundation Models: ICML 2025 Workshop (2025). https://openreview.net/forum?id=K00sxDnBBd
116. Lin, C.-Y.: Rouge: a package for automatic evaluation of summaries. In: Text Summarization Branches Out, pp. 74–81 (2004)
117. Shokri, R.: Membership inference attacks against machine learning models. In: IEEE Symposium on Security and Privacy (SP), pp. 3–18. IEEE (2017)
118. Shi, W., et al.: Detecting pretraining data from large language models. In: The Twelfth International Conference on Learning Representations (2024). https://openreview.net/forum?id=zWqr3MQuNs
119. Zou, A.: Universal and transferable adversarial attacks on aligned language models (2023). arXiv:2307.15043
120. Hendrycks, D., et al.: Measuring massive multitask language understanding. In: International Conference on Learning Representations (2021). https://openreview.net/forum?id=d7KBjmI3GmQ
121. Zheng, L., et al.: Judging llm-as-a-judge with mt-bench and chatbot arena. Adv. Neural. Inf. Process. Syst. **36**, 46595–46623 (2023)

Chapter 3
Real-Time Data Influence Measurement

Abstract Data valuation quantifies the contribution of each training data point, enabling harmful data detection and enhancing model robustness. However, existing methods are typically post-hoc and require fully trained models, making them computationally expensive and unable to detect harmful data early in training. We develop a real-time, trajectory-based data valuation method that assesses training data by analyzing their influence on the optimization path. The proposed approach includes three key innovations: a real-time valuation framework with minimal overhead that can be seamlessly integrated into standard training processes, an adaptive reference point mechanism that evaluates data impact on generalization, and a normalization technique that ensures fair comparisons across training stages. Theoretical analysis shows that the method achieves directional alignment, boundedness, stability, and fairness. Experiments demonstrate that our approach achieves up to 180× speedup over baseline methods while maintaining robust performance across diverse models and datasets.

Motivation and Contribution

Training data points contribute differently to model learning. Some enhance generalization, while others introduce harmful noise that degrades performance. Data valuation, which quantifies the contribution of each data point, is critical for robust and efficient machine learning [1, 2].

Traditional data valuation methods, including Leave-One-Out (LOO), Shapley Value [3–5], and influence functions [6, 7], measure data value by how test performance changes when a data point is removed. While intuitive, these post-hoc approaches have fundamental limitations. First, these methods evaluate data influence only on the final trained models, preventing early detection of harmful training data; when harmful data can be identified, it may have already corrupted model optimization irreversibly [8, 9]. Second, they often misjudge high-loss data as low-value data, since challenging yet informative data are high-loss but high-value because they enhance generalization by enabling robust representations and precise decision boundaries [10–12]. Third, they are computationally intensive because they require

J. Xu and X. Jia, *Machine Unlearning*, Wireless Networks,
https://doi.org/10.1007/978-3-032-30978-5_3

model retraining [3], second-order approximations [6, 13], or full gradient trajectory storage [14] to evaluate each data point's influence, making them impractical for real-time or dynamic training scenarios.

To overcome these limitations, we propose an online, trajectory-aware valuation approach. Unlike existing methods, the proposed approach continuously assesses how each data point guides the optimization trajectory toward better generalization, thereby enabling early detection of harmful data without requiring convergence.

Despite its promise, trajectory-based data valuation poses significant challenges. First, it is computationally expensive to track each data point's influence over the entire training trajectory, because doing so requires storing or recomputing training data gradients at every training step, resulting in prohibitive overhead for large datasets or long training periods. Second, there is a lack of principled criteria to determine whether a training example actually drives the parameters toward a region that generalizes better, rather than merely reducing the current batch loss. Third, data used in the early training stage may receive unfairly high values, since gradient magnitudes naturally decrease during training, causing data used earlier to produce larger parameter updates and receive higher values regardless of their intrinsic quality.

To address these challenges, the proposed method develops three key innovations. First, it seamlessly integrates with standard optimization processes, eliminating expensive post-hoc recomputation. Second, it employs an adaptive reference point mechanism that provides reliable benchmarks for evaluating data contributions to optimization progress. Third, it normalizes contributions across training stages to compensate for gradient decay, ensuring fair valuation throughout learning.

The main contributions of this chapter can be summarized as follows. First, we introduce a real-time, trajectory-based data valuation framework that evaluates data contributions during training without requiring model convergence. Second, we design an adaptive reference point mechanism with stage-aware normalization that enables fair and reliable data valuation across different training stages, effectively addressing gradient decay bias. Third, we provide theoretical guarantees showing that the proposed framework produces directionally aligned, bounded, stable, and fair valuations; extensive experiments show that it achieves up to 180× speedup over baseline methods while maintaining robust performance across diverse models and datasets.

System Model and Problem Statement

Machine learning systems train models by iteratively updating parameters using mini-batches of data drawn from a training dataset. The system model formalizes this standard training pipeline together with the information available at each iteration, while the problem statement defines the real-time data valuation objective that the proposed framework must satisfy.

System Model

Consider a standard supervised learning setting with a training dataset $\mathcal{D} = \{(\mathbf{x}_i, y_i)\}_{i=1}^{N}$ consisting of N data points drawn from a distribution P. A model $f(\cdot\,; \boldsymbol{\theta})$, parameterized by $\boldsymbol{\theta} \in \mathbb{R}^d$, is trained by minimizing the empirical risk

$$\hat{R}(\boldsymbol{\theta}; \mathcal{D}) = \frac{1}{N} \sum_{i=1}^{N} \ell\big(f(\mathbf{x}_i; \boldsymbol{\theta}),\ y_i\big), \tag{3.1}$$

where $\ell(\cdot, \cdot)$ denotes the loss function.

Training proceeds via stochastic gradient descent (SGD) over T iterations. At each step t, a mini-batch $\mathcal{B}_t \subset \mathcal{D}$ of size B is sampled, and the parameters are updated as

$$\boldsymbol{\theta}_t = \boldsymbol{\theta}_{t-1} - \eta_t \frac{1}{|\mathcal{B}_t|} \sum_{i \in \mathcal{B}_t} \nabla_{\boldsymbol{\theta}} \ell\big(f(\mathbf{x}_i; \boldsymbol{\theta}_{t-1}),\ y_i\big), \tag{3.2}$$

where η_t is the learning rate at step t. The sequence of parameter states $\boldsymbol{\theta}_0, \boldsymbol{\theta}_1, \ldots, \boldsymbol{\theta}_T$ defines the *optimization trajectory* of the model.

At every iteration, the following quantities are available to the valuation system: the current parameters $\boldsymbol{\theta}_{t-1}$, the learning rate η_t, the mini-batch composition $\mathcal{B}_t$, and the per-sample gradients $\nabla_{\boldsymbol{\theta}} \ell(f(\mathbf{x}_i; \boldsymbol{\theta}_{t-1}), y_i)$ for each $i \in \mathcal{B}_t$. These quantities are produced as a byproduct of standard training, so recording them introduces minimal computational overhead.

Problem Statement

The goal is to evaluate the contribution of individual training data points to the model's learning process in real time, based on their influence on the optimization trajectory. Specifically, the method should evaluate whether a data point's gradient update moves the model toward regions in the parameter space associated with better generalization, which can be approximated using a reference point $\boldsymbol{\theta}^t_{\text{ref}}$ at each step t. Training data are considered more valuable if their gradient updates move the model parameters closer to the reference point.

Formally, for a training data point $(\mathbf{x}_i, y_i)$ included in the mini-batch $\mathcal{B}_t$ at step t, its real-time value is defined as

$$v_i^t = \mathcal{V}_t\big((\mathbf{x}_i, y_i),\ \boldsymbol{\theta}_{t-1},\ \boldsymbol{\theta}^t_{\text{ref}}\big), \tag{3.3}$$

where $\boldsymbol{\theta}^t_{\text{ref}}$ is a reference point that serves as a proxy for improved generalization, and $\mathcal{V}_t(\cdot)$ quantifies how effectively the data point guides the model toward this improved state in the parameter space.

The cumulative value of a data point across the entire training process is then defined as

$$v_i^{[0,T]} = \sum_{t=1}^{T} v_i^t, \tag{3.4}$$

which captures its total influence on model training from initialization to completion.

The method must satisfy three key requirements: (1) computational and storage efficiency that enables real-time data valuation during training without requiring model convergence or retraining, (2) cross-stage fairness that ensures consistent value comparisons across different training phases despite natural variations in gradient magnitudes and parameter sensitivity, and (3) trajectory alignment that captures each data point's contribution to the optimization trajectory.

Trajectory-Based Data Valuation Framework

Data valuation should capture how each training point dynamically steers the optimization trajectory, not merely how test loss changes when the point is removed from a fully trained model. Most existing data valuation methods [3, 5, 6] evaluate contributions by comparing model performance (e.g., test loss) with and without specific data points. These loss-based approaches suffer from several limitations, including inconsistent valuations [15], sensitivity to the choice of the loss function, high retraining costs [3], and dependence on full model convergence [6]. Most importantly, these methods provide only a static snapshot of data utility, failing to capture how data dynamically influences the optimization process. This static perspective systematically undervalues difficult-but-informative training data (e.g., rare-class or decision-boundary points) that initially cause high prediction errors but become invaluable once the model develops sufficient representational capacity [10, 11].

While recent gradient-based methods such as TracIn [14] and GraND [16] incorporate training dynamics into their valuation schemes, they operate with a local perspective. These approaches evaluate gradient similarity or magnitude at specific model states, capturing only instantaneous gradient contributions without establishing connections to long-term influence. Consequently, they fail to assess whether immediate gradient contributions align with the overall optimization objective or merely reflect transient training fluctuations.

To address these limitations, this section develops a trajectory-based data valuation framework that measures data value through its dynamic influence on the entire optimization trajectory. In this framework, training data are considered valuable if they consistently guide the model parameters toward regions of the parameter space associated with better generalization. In practice, this desirable region can be approximated by the final converged parameter $\boldsymbol{\theta}^*$, under the assumption that the optimized model reflects improved generalization capabilities. Data contributions can then be evaluated based on how effectively each training data point moves the model toward the optimal parameters. The remainder of this section formalizes this idea: section "Basic Method with Static Reference Points" introduces a basic method that uses the final model parameters as a static reference point, section "Real-Time Measurement with Adaptive Reference Points" extends it to LiveVal with adaptive reference points for real-time valuation, and section "Real-Time Implementation via Dual Queues" describes the dual-queue implementation that makes real-time evaluation practical.

Basic Method with Static Reference Points

The optimization process constitutes a trajectory in parameter space, where training data contribute through gradient computations at each iteration. Different data points exert different influences on the direction and magnitude of parameter updates, and their cumulative effects shape the overall trajectory. During SGD, the key quantities at each iteration t, namely the parameters $\boldsymbol{\theta}_{t-1}$, the learning rate η_t, and the mini-batch $\mathcal{B}_t$, are already available as part of the standard training loop described in section "System Model".

To evaluate each training data point's contribution to this trajectory, we measure how effectively it guides the optimization toward the final converged parameter $\boldsymbol{\theta}^*$. For a data point $i \in \mathcal{B}_t$, we first establish the direction toward the final model parameter:

$$\Delta\boldsymbol{\theta}_t = \boldsymbol{\theta}^* - \boldsymbol{\theta}_{t-1}, \tag{3.5}$$

which represents the ideal displacement from the current state to the converged solution.

To assess a data point's effectiveness in moving toward this target, we introduce a hypothetical residual vector $\mathbf{u}_i^t$ representing the remaining displacement to $\boldsymbol{\theta}^*$ after applying only that data point's gradient update:

$$\mathbf{u}_i^t = \boldsymbol{\theta}^* - \big(\boldsymbol{\theta}_{t-1} - \eta_t \nabla_{\boldsymbol{\theta}} \ell(f(\mathbf{x}_i; \boldsymbol{\theta}_{t-1}), y_i)\big). \tag{3.6}$$

As illustrated in Fig. 3.1, $\|\mathbf{u}_i^t\|$ captures the remaining distance to $\boldsymbol{\theta}^*$ after this hypothetical update. A reduction in distance ($\|\mathbf{u}_i^t\| < \|\Delta\boldsymbol{\theta}_t\|$) indicates that the data point moves the model closer to the target, while an increase ($\|\mathbf{u}_i^t\| > \|\Delta\boldsymbol{\theta}_t\|$) indicates that it moves the model away.

However, direct distance comparisons across different training stages are unfair because gradient magnitudes naturally decrease as training progresses. To address this, the step-wise value is defined with a scale-invariant normalization:

$$v_i^t = \frac{\|\Delta\boldsymbol{\theta}_t\| - \|\mathbf{u}_i^t\|}{\|\Delta\boldsymbol{\theta}_t\| + \|\mathbf{u}_i^t\|}. \tag{3.7}$$

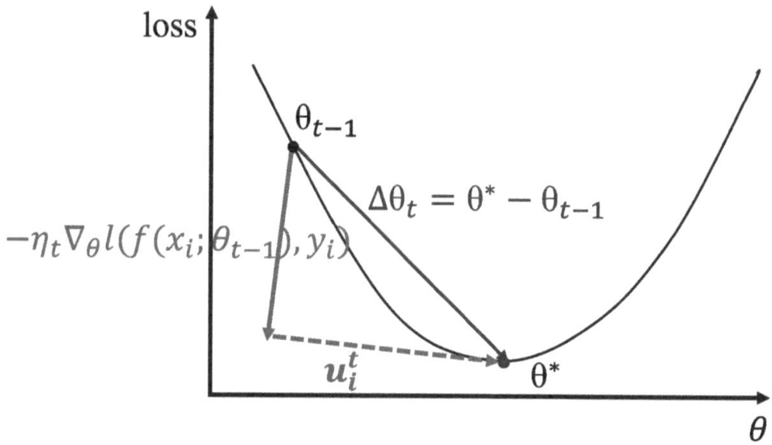

Fig. 3.1 Basic method using the final model parameter as a static reference point. The vector $\Delta\boldsymbol{\theta}_t$ points from the current parameters to $\boldsymbol{\theta}^*$, and $\mathbf{u}_i^t$ is the residual after a hypothetical single-sample gradient step

The numerator $\|\Delta\boldsymbol{\theta}_t\| - \|\mathbf{u}_i^t\|$ measures the improvement toward the reference point: a positive value indicates that the data point helps guide the model toward $\boldsymbol{\theta}^*$, while a negative value indicates that it pushes the model away. The denominator $\|\Delta\boldsymbol{\theta}_t\| + \|\mathbf{u}_i^t\|$ provides scale normalization, bounding the step-wise value within $[-1, 1]$ and thereby enabling fair comparison across training stages. A detailed theoretical analysis of these properties is presented in section "Theoretical Analysis".

The cumulative value over the entire training run is obtained by summing step-wise values in accordance with Eq. (3.4), with the convention that $v_i^t = 0$ whenever $i \notin \mathcal{B}_t$. This cumulative score captures each data point's total contribution throughout training.

Real-Time Measurement with Adaptive Reference Points

The basic method defined in section "Basic Method with Static Reference Points" relies on the final model parameters for valuation, making it a post-hoc procedure unsuitable for real-time analysis. Moreover, it requires storing the entire optimization trajectory, which incurs significant storage overhead. To address these limitations, we propose LiveVal, a real-time valuation method that employs an adaptive reference point mechanism. In this context, "real-time" refers to the ability to provide data influence estimates during training, enabling early harmful data detection without waiting for full convergence. Instead of using the final parameter $\boldsymbol{\theta}^*$, LiveVal adaptively selects near-future model parameters as reference points, thereby enabling valuation as training unfolds.

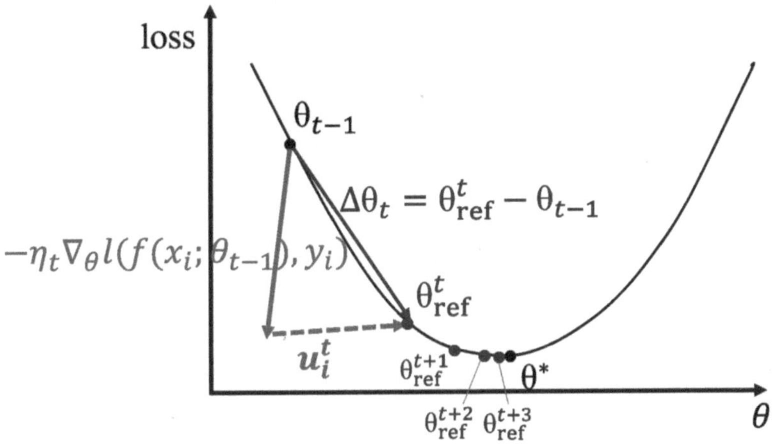

Fig. 3.2 LiveVal's adaptive reference point mechanism. The look-ahead window δ_t expands when loss changes rapidly and contracts as training converges, producing reference points that balance stability and temporal resolution

The adaptive reference point mechanism comprises two core components. Reference point selection determines which specific future parameter $\boldsymbol{\theta}_{t-1+\delta_t}$ to use at each step, while dynamic window adjustment optimizes the look-ahead window size δ_t, expanding it during steep gradient descent and shrinking it near convergence, as illustrated in Fig. 3.2.

Reference Point Selection

At each step t, the reference point $\boldsymbol{\theta}^t_{\text{ref}}$ is set to a future model parameter:

$$\boldsymbol{\theta}^t_{\text{ref}} = \begin{cases} \boldsymbol{\theta}_{t-1+\delta_t} & \text{if } t - 1 + \delta_t < T, \\ \boldsymbol{\theta}_T & \text{otherwise,} \end{cases} \tag{3.8}$$

where δ_t is the look-ahead window size and T is the total number of training steps. This dynamic selection avoids waiting for full convergence while maintaining storage efficiency, since only the parameters within the current window need to be retained. The key challenge lies in adapting δ_t at each training step to balance evaluation accuracy with computational efficiency.

Dynamic Window Adjustment

A fixed window size cannot adapt to the varying dynamics of model training. Early training requires longer windows to smooth noisy updates, while near convergence shorter windows capture subtle progress with higher temporal resolution. LiveVal adjusts the look-ahead window δ_t at each step t based on the loss change rate $\dot{L}$:

$$\dot{L} = \frac{L_{t'} - L_{t-1}}{\delta_{t-1}}, \tag{3.9}$$

where L_{t-1} is the average training loss at step $t-1$ and $L_{t'}$ is the loss at the future step t':

$$L_{t-1} = \frac{1}{|\mathcal{B}_{t-1}|} \sum_{i \in \mathcal{B}_{t-1}} \ell\big(f(\mathbf{x}_i; \boldsymbol{\theta}_{t-1}), y_i\big), \tag{3.10}$$

$$L_{t'} = \frac{1}{|\mathcal{B}_{t'}|} \sum_{i \in \mathcal{B}_{t'}} \ell\big(f(\mathbf{x}_i; \boldsymbol{\theta}_{t'}), y_i\big). \tag{3.11}$$

Here $t' = \min(t - 1 + \delta_{t-1},\ T)$ denotes the future step used for computing the rate of loss change, bounded by the total training duration T.

Based on the magnitude of $\dot{L}$, the window size is updated as:

$$\delta_t = \begin{cases} \min(\delta_{t-1} + \Delta\delta,\ \delta_{\max}) & \text{if } |\dot{L}| > \varepsilon_{\max}, \\ \max(\delta_{t-1} - \Delta\delta,\ \delta_{\min}) & \text{if } |\dot{L}| < \varepsilon_{\min}, \\ \delta_{t-1} & \text{otherwise,} \end{cases} \tag{3.12}$$

where $\Delta\delta$ controls the adaptation rate, $\delta_{\min}$ and $\delta_{\max}$ bound the look-ahead window size, and the thresholds $\varepsilon_{\min}$ and $\varepsilon_{\max}$ trigger adjustments. When the loss changes rapidly ($|\dot{L}| > \varepsilon_{\max}$), indicating volatile training dynamics, the window is enlarged to obtain more stable reference points and smooth out short-term fluctuations. Conversely, when the loss stabilizes near convergence ($|\dot{L}| < \varepsilon_{\min}$), the window shrinks to capture fine-grained progress with higher temporal resolution. When the rate of loss change falls between these two thresholds, the window size remains unchanged, maintaining a balance between sensitivity to local fluctuations and awareness of longer-term optimization trends.

LiveVal Data Value Computation

With the adaptively selected reference point $\boldsymbol{\theta}^t_{\text{ref}}$, LiveVal computes the step-wise data value by substituting the adaptive reference into the valuation formula from Eq. (3.7). Specifically, the direction vector, residual, and normalized value are computed as:

$$\Delta\boldsymbol{\theta}_t = \boldsymbol{\theta}^t_{\text{ref}} - \boldsymbol{\theta}_{t-1}, \tag{3.13}$$

$$\mathbf{u}^t_i = \boldsymbol{\theta}^t_{\text{ref}} - \big(\boldsymbol{\theta}_{t-1} - \eta_t \nabla_{\boldsymbol{\theta}} \ell(f(\mathbf{x}_i; \boldsymbol{\theta}_{t-1}), y_i)\big), \tag{3.14}$$

$$v^t_i = \frac{\|\Delta\boldsymbol{\theta}_t\| - \|\mathbf{u}^t_i\|}{\|\Delta\boldsymbol{\theta}_t\| + \|\mathbf{u}^t_i\|}. \tag{3.15}$$

The cumulative value $v_i^{[0,T]}$ is obtained by summing step-wise values as in Eq. (3.4). The only structural difference from the basic method is that $\boldsymbol{\theta}^t_{\text{ref}}$ is now a near-future parameter selected adaptively via Eqs. (3.8) and (3.12), rather than the fixed final parameter $\boldsymbol{\theta}^*$.

Real-Time Implementation via Dual Queues

LiveVal is implemented using a dual-queue architecture (Algorithm 1) that enables real-time valuation during training without storing complete optimization trajectories.

The system maintains two synchronized queues. The Model Queue Q_θ stores recent training states $(t, \boldsymbol{\theta}_t, \mathcal{B}_t, \eta_t, L_t)$ within a sliding window $[t - \delta_{\max}, t]$, retaining only the information needed for pending valuations. The Reference Queue Q_{ref} tracks pending evaluations as $(t_{\text{eval}}, t_{\text{ref}})$ pairs, where t_{eval} is the step whose data are to be valued and t_{ref} is the step that supplies the reference parameters.

During each training step t, LiveVal checks whether the current parameters $\boldsymbol{\theta}_t$ serve as reference points for any pending evaluations in Q_{ref}. When matches are found, it computes data values for all training data points in the corresponding historical batches using Eqs. (3.13)–(3.15). Processed entries are then removed from both queues to maintain storage efficiency.

This design reduces storage from $O(Td)$ for full trajectory storage to $O(\delta_{\max} d)$, where d is the parameter dimension and $\delta_{\max}$ is the maximum adaptive window size. The computational complexity is $O(TBd)$ over T training steps with batch size B, matching the cost of standard training up to a constant factor. Unlike existing methods that require model retraining or post-hoc analysis, LiveVal provides data influence estimates as training proceeds, enabling early detection of harmful data before their effects propagate deeply into the model parameters.

Algorithm 1 LiveVal: Real-time Data Valuation

Require: Training dataset $\{(\mathbf{x}_i, y_i)\}_{i=1}^{N}$, Total steps T, Initial window δ_0, Window bounds $\delta_{\min}$, $\delta_{\max}$, Adaptation rate $\Delta\delta$, Loss thresholds $\varepsilon_{\min}$, $\varepsilon_{\max}$
Ensure: Step-wise values $\{v_i^t\}_{t=1}^{T}$, Cumulative value $v_i^{[0,T]}$

Initialize $v_i^{[0,T]} \leftarrow 0$ for all i
Initialize array $\{v_i^t\}_{t=1}^{T} \leftarrow \mathbf{0}$ for all i
Initialize $\boldsymbol{\theta}_0$ and L_0
Initialize window size array $\{\delta_t\}_{t=1}^{T} \leftarrow \{\delta_0\}$
Initialize model queue Q_θ for storing model parameters
Initialize reference queue Q_{ref} for storing $(t_{\text{eval}}, t_{\text{ref}})$ pairs
for $t = 1$ to T **do**
  Sample mini-batch $\mathcal{B}_t$
  $\mathbf{g}_t \leftarrow \frac{1}{|\mathcal{B}_t|}\sum_{i\in\mathcal{B}_t} \nabla_{\boldsymbol{\theta}}\ell(f(\mathbf{x}_i;\boldsymbol{\theta}_{t-1}), y_i)$
  $\boldsymbol{\theta}_t \leftarrow \boldsymbol{\theta}_{t-1} - \eta_t \mathbf{g}_t$
  $L_t \leftarrow \frac{1}{|\mathcal{B}_t|}\sum_{i\in\mathcal{B}_t} \ell(f(\mathbf{x}_i;\boldsymbol{\theta}_t), y_i)$
  Store $(t, \boldsymbol{\theta}_t, \mathcal{B}_t, \eta_t, L_t)$ in Q_θ
  $t_{\text{ref}} \leftarrow \min(t - 1 + \delta_{t-1}, T)$
  Add (t, t_{ref}) to Q_{ref}
  $\dot{L} \leftarrow (L_t - L_{t-1})/\delta_{t-1}$
  if $|\dot{L}| > \varepsilon_{\max}$ **then**
    $\delta_t \leftarrow \min(\delta_{t-1} + \Delta\delta, \delta_{\max})$
  else if $|\dot{L}| < \varepsilon_{\min}$ **then**
    $\delta_t \leftarrow \max(\delta_{t-1} - \Delta\delta, \delta_{\min})$
  else
    $\delta_t \leftarrow \delta_{t-1}$
  end if
  if $\exists(t', t) \in Q_{\text{ref}}$ **then**
    Get all pairs (t', t) from Q_{ref}
    for each (t', t) pair **do**
      Get $\boldsymbol{\theta}_{t'-1}$ from Q_θ
      $\Delta\boldsymbol{\theta}_{t'} \leftarrow \boldsymbol{\theta}_t - \boldsymbol{\theta}_{t'-1}$
      for each $i \in \mathcal{B}_{t'}$ **do**
        $\mathbf{g}_i \leftarrow \nabla_{\boldsymbol{\theta}}\ell(f(\mathbf{x}_i;\boldsymbol{\theta}_{t'-1}), y_i)$
        $\mathbf{u}_i^{t'} \leftarrow \boldsymbol{\theta}_t - (\boldsymbol{\theta}_{t'-1} - \eta_{t'}\mathbf{g}_i)$
        $v_i^{t'} \leftarrow \frac{\|\Delta\boldsymbol{\theta}_{t'}\| - \|\mathbf{u}_i^{t'}\|}{\|\Delta\boldsymbol{\theta}_{t'}\| + \|\mathbf{u}_i^{t'}\|}$
        $v_i^{[0,T]} \leftarrow v_i^{[0,T]} + v_i^{t'}$
      end for
      Remove (t', t) from Q_{ref}
    end for
  end if
  Remove from Q_θ where step $< t - \delta_{\max}$
  Remove from Q_{ref} where $t_{\text{ref}} < t$
end for
return $\{v_i^t\}_{t=1}^{T}$, $v_i^{[0,T]}$

Theoretical Analysis

LiveVal's valuation formula (Eq. (3.15)) is grounded in three theoretical properties: reliability (directional alignment and boundedness), stability under stochastic training, and temporal fairness. This section formalizes each property under standard optimization assumptions, providing guarantees that the valuations produced by LiveVal are meaningful, consistent, and unbiased across training stages.

Assumptions and Fundamental Properties

The theoretical results in this section rely on three standard assumptions from stochastic optimization theory.

Assumption 1 (Smoothness) *The loss function* $\ell(f(\mathbf{x};\boldsymbol{\theta}), y)$ *is* β*-smooth, i.e., for all* $\boldsymbol{\theta}, \boldsymbol{\theta}' \in \mathbb{R}^d$,

$$\|\nabla_{\boldsymbol{\theta}}\ell(f(\mathbf{x};\boldsymbol{\theta}), y) - \nabla_{\boldsymbol{\theta}}\ell(f(\mathbf{x};\boldsymbol{\theta}'), y)\| \leq \beta\|\boldsymbol{\theta} - \boldsymbol{\theta}'\|. \tag{3.16}$$

Assumption 2 (Bounded gradients) *Stochastic gradients are clipped and bounded by a constant* $G > 0$[1]*:*

$$\|\nabla_{\boldsymbol{\theta}}\ell(f(\mathbf{x};\boldsymbol{\theta}), y)\| \leq G. \tag{3.17}$$

Assumption 3 (Stable step size) *For every iteration* t*, the learning rate* η_t *satisfies*

$$\eta_t G \leq \|\Delta\boldsymbol{\theta}_t\|. \tag{3.18}$$

Assumption 1 ensures that gradients vary continuously, which is needed for the perturbation analysis in section "Stability Analysis". Assumption 2 bounds the magnitude of any single gradient update, a condition that holds automatically when gradient clipping is applied. Assumption 3 requires that the learning-rate-scaled gradient is small relative to the distance to the reference point, ensuring that individual updates do not overshoot the reference. Together, these three assumptions underpin the reliability, stability, and fairness guarantees established below.

Under these assumptions, the step-wise data value defined in Eq. (3.15) satisfies two fundamental properties: directional alignment and value boundedness.

Theorem 3.1 (Fundamental Properties of LiveVal) *For any training data point* i *and iteration* t*, the step-wise data value* v_i^t *satisfies:*

1. ***Directional Alignment.*** *If the gradient update moves the parameters closer to the reference point* $\boldsymbol{\theta}^t_{ref}$*, then* $v_i^t \geq 0$.
2. ***Value Boundedness.*** *For any training data point* i *and any iteration* t*,* $v_i^t \in [-1, 1]$.

[1] Gradient clipping is standard practice in deep networks to prevent exploding gradients [17].

Proof *Part 1 (Directional Alignment).* If the gradient update moves the parameters closer to $\boldsymbol{\theta}^t_{\text{ref}}$, then by definition

$$\|\boldsymbol{\theta}^t_{\text{ref}} - (\boldsymbol{\theta}_{t-1} - \eta_t \nabla_{\boldsymbol{\theta}} \ell(f(\mathbf{x}_i; \boldsymbol{\theta}_{t-1}), y_i))\| \leq \|\boldsymbol{\theta}^t_{\text{ref}} - \boldsymbol{\theta}_{t-1}\|. \tag{3.19}$$

Substituting the definitions from Eqs. (3.13) and (3.14), Eq. (3.19) is equivalent to $\|\mathbf{u}^t_i\| \leq \|\Delta\boldsymbol{\theta}_t\|$, which implies $\|\Delta\boldsymbol{\theta}_t\| - \|\mathbf{u}^t_i\| \geq 0$. Since the denominator $\|\Delta\boldsymbol{\theta}_t\| + \|\mathbf{u}^t_i\|$ is strictly positive, it follows that $v^t_i \geq 0$.

Part 2 (Value Boundedness). Since $\|\mathbf{u}^t_i\| \geq 0$ and $\|\Delta\boldsymbol{\theta}_t\| \geq 0$, the following two inequalities hold:

$$\|\Delta\boldsymbol{\theta}_t\| - \|\mathbf{u}^t_i\| \leq \|\Delta\boldsymbol{\theta}_t\| + \|\mathbf{u}^t_i\| \implies v^t_i \leq 1, \tag{3.20}$$

$$\|\mathbf{u}^t_i\| - \|\Delta\boldsymbol{\theta}_t\| \leq \|\Delta\boldsymbol{\theta}_t\| + \|\mathbf{u}^t_i\| \implies v^t_i \geq -1. \tag{3.21}$$

Therefore $v^t_i \in [-1, 1]$. □

Together, these two properties ensure that LiveVal assigns positive values to gradient updates that move the model toward the reference point and negative values to updates that move it away, while maintaining a consistent $[-1, 1]$ scale that is independent of the absolute magnitudes of gradients or parameter displacements. This bounded, sign-consistent behavior is the foundation on which the stability and fairness results in the following subsections build.

Stability Analysis

Stochastic training introduces randomness through mini-batch sampling and parameter initialization, and a practical valuation method must produce stable estimates despite this randomness. This subsection quantifies the volatility of LiveVal's stepwise values and shows that it decreases with the learning rate, ensuring increasingly reliable valuations as training progresses.

Definition 3.1 (*Data Valuation Volatility*) The volatility of data valuation v^t_i for training data point i at step t is

$$\sigma^{[t]}_i = \sqrt{\mathbb{E}\Big[\big(v^t_i - \mathbb{E}[v^t_i]\big)^2\Big]}, \tag{3.22}$$

where the expectation is over all training randomness (mini-batch sampling and initialization).

Theorem 3.2 (Stability under Stochastic Training) *Under Assumptions 1 through 3 and the condition* $\|\Delta\boldsymbol{\theta}_t\| \geq D_{\min} > 0$*, the volatility of* v_i^t *is bounded by*

$$\sigma_i^{[t]} \leq \frac{\eta_t G}{\|\Delta\boldsymbol{\theta}_t\|} \leq \frac{\eta_t G}{D_{\min}}. \tag{3.23}$$

Proof *Step 1: Bound the effect of a gradient update on parameter distance.* From the definitions in Eqs. (3.13) and (3.14), the residual vector can be written as $\mathbf{u}_i^t = \Delta\boldsymbol{\theta}_t + \eta_t \nabla_{\boldsymbol{\theta}} \ell(f(\mathbf{x}_i; \boldsymbol{\theta}_{t-1}), y_i)$. Applying the reverse triangle inequality and the bounded-gradient assumption (Assumption 2),

$$\big|\|\mathbf{u}_i^t\| - \|\Delta\boldsymbol{\theta}_t\|\big| \leq \|\mathbf{u}_i^t - \Delta\boldsymbol{\theta}_t\| = \eta_t \big\|\nabla_{\boldsymbol{\theta}} \ell(f(\mathbf{x}_i; \boldsymbol{\theta}_{t-1}), y_i)\big\| \leq \eta_t G. \tag{3.24}$$

Step 2: Lower-bound the denominator. By the triangle inequality,

$$\begin{aligned} \|\mathbf{u}_i^t\| &= \|\Delta\boldsymbol{\theta}_t + \eta_t \nabla_{\boldsymbol{\theta}} \ell(f(\mathbf{x}_i; \boldsymbol{\theta}_{t-1}), y_i)\| \\ &\geq \|\Delta\boldsymbol{\theta}_t\| - \eta_t \|\nabla_{\boldsymbol{\theta}} \ell(f(\mathbf{x}_i; \boldsymbol{\theta}_{t-1}), y_i)\| \geq \|\Delta\boldsymbol{\theta}_t\| - \eta_t G. \end{aligned} \tag{3.25}$$

Therefore $\|\Delta\boldsymbol{\theta}_t\| + \|\mathbf{u}_i^t\| \geq 2\|\Delta\boldsymbol{\theta}_t\| - \eta_t G$. Under Assumption 3 ($\eta_t G \leq \|\Delta\boldsymbol{\theta}_t\|$), this yields $\|\Delta\boldsymbol{\theta}_t\| + \|\mathbf{u}_i^t\| \geq \|\Delta\boldsymbol{\theta}_t\|$. Combining with the result of Step 1,

$$|v_i^t| = \frac{\big|\|\Delta\boldsymbol{\theta}_t\| - \|\mathbf{u}_i^t\|\big|}{\|\Delta\boldsymbol{\theta}_t\| + \|\mathbf{u}_i^t\|} \leq \frac{\eta_t G}{\|\Delta\boldsymbol{\theta}_t\|}. \tag{3.26}$$

Step 3: Bound the volatility via variance decomposition. Since the variance satisfies

$$\mathrm{Var}(v_i^t) = \mathbb{E}\big[(v_i^t)^2\big] - \big(\mathbb{E}[v_i^t]\big)^2 \leq \mathbb{E}\big[(v_i^t)^2\big], \tag{3.27}$$

the bound from Step 2 gives

$$\sigma_i^{[t]} = \sqrt{\mathrm{Var}(v_i^t)} \leq \sqrt{\mathbb{E}\big[(v_i^t)^2\big]} \leq \frac{\eta_t G}{\|\Delta\boldsymbol{\theta}_t\|}. \tag{3.28}$$

□

The bound in Eq. (3.23) shows that valuation volatility decreases proportionally with the learning rate η_t. In standard training schedules where the learning rate decays over time, LiveVal's valuations therefore become increasingly reliable as training progresses. The additional condition $\|\Delta\boldsymbol{\theta}_t\| \geq D_{\min}$ ensures that the reference point is not trivially close to the current parameters, which would make the normalization ill-conditioned.

The stability theorem guarantees that LiveVal's valuations are robust to training stochasticity, with volatility that shrinks as the learning rate decays. The next subsection addresses a complementary concern: whether data points appearing at different stages of training receive systematically biased valuations.

Fairness Analysis

Traditional gradient-based valuation methods often exhibit temporal bias: data encountered early in training, when gradient magnitudes are large, receive disproportionately high values regardless of their intrinsic quality, while late-stage contributions are systematically undervalued. Temporal fairness requires that data points with equivalent optimization contributions receive equal valuations irrespective of when they appear in training.

Definition 3.2 (*Temporal Fairness*) A data valuation method satisfies temporal fairness if training data with equivalent optimization contributions receive equal valuations regardless of their temporal position, given identical gradients and relative orientations to reference vectors.

Theorem 3.3 (Temporal Fairness of LiveVal) *Let training data points i and j be used at steps t_1 and t_2, respectively. If the following four conditions hold:*

1. *equal gradients: $\nabla_{\boldsymbol{\theta}}\ell(f(\mathbf{x}_i;\boldsymbol{\theta}_{t_1-1}), y_i) = \nabla_{\boldsymbol{\theta}}\ell(f(\mathbf{x}_j;\boldsymbol{\theta}_{t_2-1}), y_j) = \mathbf{g}$;*
2. *equal learning rates: $\eta_{t_1} = \eta_{t_2} = \eta$;*
3. *equal reference distances: $\|\boldsymbol{\theta}_{ref}^{t_1} - \boldsymbol{\theta}_{t_1-1}\| = \|\boldsymbol{\theta}_{ref}^{t_2} - \boldsymbol{\theta}_{t_2-1}\| = D$;*
4. *equal directional alignment: $\frac{\langle \Delta\boldsymbol{\theta}_{t_1}, \mathbf{g}\rangle}{\|\Delta\boldsymbol{\theta}_{t_1}\|\,\|\mathbf{g}\|} = \frac{\langle \Delta\boldsymbol{\theta}_{t_2}, \mathbf{g}\rangle}{\|\Delta\boldsymbol{\theta}_{t_2}\|\,\|\mathbf{g}\|}$;*

then LiveVal assigns equal valuations: $v_i^{t_1} = v_j^{t_2}$.

Proof From the definition of the residual vector (Eq. (3.14)) and the valuation formula (Eq. (3.15)), the residuals at the two steps are

$$\mathbf{u}_i^{t_1} = \Delta\boldsymbol{\theta}_{t_1} + \eta\mathbf{g}, \qquad \mathbf{u}_j^{t_2} = \Delta\boldsymbol{\theta}_{t_2} + \eta\mathbf{g}. \tag{3.29}$$

Expanding the squared norms via the law of cosines,

$$\|\mathbf{u}_i^{t_1}\|^2 = D^2 + \eta^2\|\mathbf{g}\|^2 + 2\eta\langle \Delta\boldsymbol{\theta}_{t_1}, \mathbf{g}\rangle, \tag{3.30}$$

$$\|\mathbf{u}_j^{t_2}\|^2 = D^2 + \eta^2\|\mathbf{g}\|^2 + 2\eta\langle \Delta\boldsymbol{\theta}_{t_2}, \mathbf{g}\rangle. \tag{3.31}$$

Condition 3 gives $\|\Delta\boldsymbol{\theta}_{t_1}\| = \|\Delta\boldsymbol{\theta}_{t_2}\| = D$, and condition 4 then implies $\langle \Delta\boldsymbol{\theta}_{t_1}, \mathbf{g}\rangle = \langle \Delta\boldsymbol{\theta}_{t_2}, \mathbf{g}\rangle$. Substituting into Eqs. (3.30) and (3.31) yields $\|\mathbf{u}_i^{t_1}\| = \|\mathbf{u}_j^{t_2}\|$, from which it follows that

$$v_i^{t_1} = \frac{D - \|\mathbf{u}_i^{t_1}\|}{D + \|\mathbf{u}_i^{t_1}\|} = \frac{D - \|\mathbf{u}_j^{t_2}\|}{D + \|\mathbf{u}_j^{t_2}\|} = v_j^{t_2}. \tag{3.32}$$

□

The temporal fairness theorem confirms that LiveVal's normalization (Eq. (3.15)) eliminates the systematic bias present in unnormalized gradient-based methods. Two data points that exert the same directional influence relative to their respective reference points receive identical valuations, regardless of whether they appear in the first or the last epoch. This property is a direct consequence of the scale-invariant denominator $\|\Delta\boldsymbol{\theta}_t\| + \|\mathbf{u}_i^t\|$, which absorbs the magnitude differences that would otherwise favor early-training data.

The three theoretical properties established in this section, namely directional alignment with boundedness, stability, and temporal fairness, collectively guarantee that LiveVal produces reliable, robust, and unbiased data valuations throughout the training process. Section "Experimental Evaluation" validates these theoretical properties empirically and evaluates LiveVal's practical effectiveness across diverse datasets, models, and corruption scenarios.

Experimental Evaluation

LiveVal is evaluated across three dimensions: accuracy and efficiency in identifying corrupted data (sections "Comparison Against Baselines" and "Early Detection Capability"), robustness across data qualities and modalities (section "Robustness Analysis"), and scalability to large models and sensitivity to hyperparameters (section "Hyperparameter Sensitivity Analysis").

Experimental Setup

Datasets and Models

The experiments span three data modalities: tabular data (Adult [18]), text data (20 Newsgroups [19]), and image data (MNIST [20]). Three models of increasing complexity are used: a Deep Neural Network (DNN), a Convolutional Neural Network (CNN, LeNet 5), and Inception V3 (approximately 24M parameters). Full implementation details are available at https://github.com/liveval/liveval.

Baselines

LiveVal is compared against five representative data valuation methods. *Leave-One-Out* (LOO) measures each data point's marginal contribution by retraining the model without that point, requiring N full retraining runs and thus incurring prohibitive computational cost. *Influence Functions* (IF) [6] approximates LOO through the inverse-Hessian-weighted gradient product:

$$v_i^{\mathrm{IF}} = -\nabla_{\boldsymbol{\theta}}\ell(f(\mathbf{z}_{\mathrm{test}}; \boldsymbol{\theta}^*), y_{\mathrm{test}})^\top H_{\boldsymbol{\theta}^*}^{-1} \nabla_{\boldsymbol{\theta}}\ell(f(\mathbf{x}_i; \boldsymbol{\theta}^*), y_i), \tag{3.33}$$

where H_{θ^*} is the Hessian matrix at the converged parameters $\boldsymbol{\theta}^*$. *GraND* [16] is a real-time method that scores training data by their expected gradient norm:

$$v_i^{\text{GraND}} = \mathbb{E}_{\boldsymbol{\theta}_t}\big[\|\nabla_{\boldsymbol{\theta}}\ell(f(\mathbf{x}_i;\boldsymbol{\theta}_t), y_i)\|_2\big]. \tag{3.34}$$

TracIn [14] aggregates gradient alignment between training and test data across checkpoints $\mathcal{T}$:

$$v_i^{\text{TracIn}} = \sum_{t\in\mathcal{T}} \eta_t \nabla_{\boldsymbol{\theta}}\ell(f(\mathbf{z}_{\text{test}};\boldsymbol{\theta}_t), y_{\text{test}})^\top \nabla_{\boldsymbol{\theta}}\ell(f(\mathbf{x}_i;\boldsymbol{\theta}_t), y_i). \tag{3.35}$$

Finally, *StaticRef* is the basic trajectory method introduced in section "Basic Method with Static Reference Points", which measures optimization guidance toward the final parameters $\boldsymbol{\theta}^*$ and serves as an ablation baseline for LiveVal's adaptive reference mechanism.

Evaluation Metrics

Following standard practice, corrupted data detection serves as a proxy task for evaluating data valuation quality. Three metrics are reported: detection accuracy (the number of truly corrupted data points among the lowest-valued data), computational efficiency (wall-clock time), and storage requirements (peak GPU memory consumption measured via `nvidia-smi`). All experiments use PyTorch 2.0 on NVIDIA RTX 2080 Ti GPUs and are repeated with eight random seeds.

Comparison Against Baselines

Theoretical Comparison

Table 3.1 summarizes the complexity and characteristics of each method. LOO requires N retraining runs with $O(N \cdot T \cdot B \cdot d)$ time complexity and $O(N \cdot d)$ storage. IF avoids retraining but relies on expensive $O(d^3)$ Hessian computations. GraND achieves computational efficiency but provides only a local gradient view without trajectory awareness. TracIn captures training dynamics but requires storing the full gradient trajectory at $O(T \cdot d)$ cost. LiveVal matches GraND's $O(T \cdot B \cdot d)$ time complexity while reducing storage to $O(\delta_{\max} \cdot d)$ and providing real-time analysis with temporal fairness guarantees (Theorem 3.3).

Table 3.1 Comparison of data valuation methods in terms of complexity and characteristics

Method	Time complexity	Storage somplexity	Real-time	Key advantages	Main limitations
LOO	$O(N \cdot T \cdot B \cdot d)$	$O(N \cdot d)$	No	Direct influence measure	Requires N retraining runs
IF	$O(d^3 + N \cdot d^2)$	$O(d^2)$	No	Efficient LOO approximation	Restrictive assumptions
GraND	$O(T \cdot B \cdot d)$	$O(d + N)$	Yes	Real-time evaluation	Local gradient focus
TracIn	$O(M \cdot T \cdot d)$	$O(T \cdot d)$	No	Training dynamics capture	Storage intensive
StaticRef	$O(T \cdot B \cdot d)$	$O(T \cdot d)$	No	Trajectory-based insights	Post-hoc analysis
LiveVal	$O(T \cdot B \cdot d)$	$O(\delta_{\max} \cdot d)$	Yes	Real-time, temporal fairness	Window size selection

Notation: N = dataset size, T = training steps, B = batch size, M = test set size, d = parameter dimension, $\delta_{\max}$ = maximum window size

Experimental Comparison

The experimental comparison evaluates detection accuracy and resource efficiency on a label-flipping detection task using MNIST.

Experimental Design

We select the MNIST dataset and flip k labels ($k \in \{10, 20, 30, 40\}$) from digit '1' to digit '7', chosen for their structural similarity, to create a challenging detection scenario. After training a CNN on the corrupted dataset, we compute data values for 100 data points (k corrupted and $100 - k$ randomly selected clean data) and count the number of flipped data points among the k lowest-valued data. Computational time and GPU memory usage are recorded during the entire valuation process.

Results and Analysis

Tables 3.2 and 3.3 reveal a clear accuracy-efficiency trade-off among the data valuation methods. LOO achieves the highest detection accuracy, identifying 8.4 out of 10 corrupted data points at the lowest corruption level, but requires prohibitive

Table 3.2 Comparison of corruption detection performance (number of corrupted data correctly identified among the k lowest-valued)

# Flipped	LOO	IF	GraND	TracIn	StaticRef	LiveVal
10	**8.4 ± 2.4**	4.6 ± 2.1	0.2 ± 0.4	4.8 ± 2.4	7.4 ± 1.3	5.4 ± 2.6
20	17.4 ± 1.7	11.8 ± 4.4	0.6 ± 0.9	**17.5 ± 2.8**	14.2 ± 1.1	12.6 ± 4.6
30	**27.0 ± 1.4**	18.0 ± 4.0	0.8 ± 1.3	26.5 ± 3.3	21.1 ± 3.7	21.3 ± 5.4
40	**37.2 ± 2.2**	24.2 ± 4.0	1.0 ± 0.7	34.1 ± 4.5	27.4 ± 1.5	31.2 ± 2.1

Table 3.3 Comparison of computational resource requirements (time and GPU memory)

Metric	LOO	IF	GraND	TracIn	StaticRef	LiveVal
Computation time	3h 34 m 29 s	1 m 29 s	13 m 2 s	7 m 42 s	2 m 14 s	**1 m 11 s**
GPU memory usage	6.817 GB	0.522 GB	0.448 GB	1.88 GB	1.24 GB	**0.521 GB**

computational resources: over 3.5 h of execution time and 6.817 GB of GPU memory. TracIn demonstrates competitive detection performance but incurs substantial overhead (7 m 42 s and 1.88 GB).

LiveVal strikes an effective balance between detection capability and computational efficiency. While it achieves slightly lower detection accuracy than LOO and TracIn, it enables early detection during training and dramatically reduces computational requirements to just 1 m 11 s and 0.521 GB of memory, representing a 180× speedup over LOO. Influence functions, by contrast, can only detect harmful data after training completion; by that point, malicious data have already corrupted the training process, wasting both time and computational resources. GraND fails at the fundamental detection task, identifying at most 1.0 corrupted data point across all corruption levels, confirming that local gradient magnitude alone is insufficient for reliable data valuation.

The baseline comparison confirms that LiveVal achieves the best efficiency among all methods with competitive detection accuracy. The next subsection examines whether LiveVal's real-time capability translates into effective early detection before training completes.

Early Detection Capability

Among all methods evaluated, only LiveVal and GraND support real-time valuation (Table 3.1). Since the baseline comparison reveals GraND's poor detection performance, LiveVal is the only effective method for early harmful data detection. This subsection analyzes how LiveVal's detection accuracy evolves across training epochs.

Figure 3.3 demonstrates LiveVal's strong early detection capability. In the $k = 40$ scenario, LiveVal identifies 11 out of 40 corrupted data points (27.5%) within the first epoch and reaches 29 (72.5%) by epoch 3. This rapid identification enables model owners to intervene during training, removing or reweighting harmful data before they significantly impact model convergence, which validates LiveVal's practical utility for enhancing model robustness.

LiveVal detects the majority of corrupted data within the first few training epochs, well before convergence. The next subsection evaluates whether this detection capability generalizes across different corruption types, data modalities, and model scales.

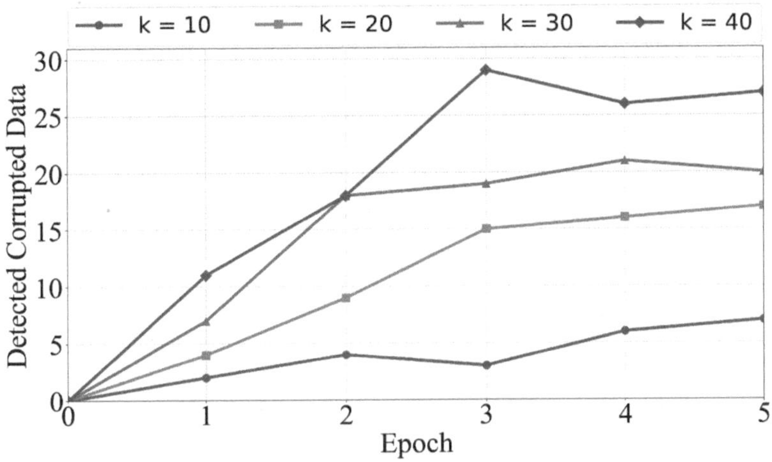

Fig. 3.3 Early detection performance of LiveVal across training epochs. The curves show the number of identified corrupted data points in the first five epochs for different corruption levels ($k = 10, 20, 30, 40$)

Robustness Analysis

The preceding experiments focus on label flipping in image data. A practical data valuation method must also handle feature-level corruption, generalize across data modalities, and scale to large model architectures. This subsection evaluates LiveVal along all three dimensions.

Robustness to Feature Corruption

Experimental Design

Using the MNIST dataset with a LeNet-5 model, we inject Gaussian noise with standard deviations $\sigma \in \{1.0, 2.0, 5.0\}$ into k randomly selected data points ($k \in \{10, 20, 30, 40\}$). After training on this modified dataset, we compute data values using LiveVal and evaluate detection performance by counting how many corrupted data points appear among the k lowest-valued data within a pool of 100 (all k corrupted plus $100 - k$ clean data).

Results and Analysis

Table 3.4 shows the number of corrupted data points successfully detected. LiveVal exhibits strong adaptability to different levels of feature corruption, with detection rates increasing as noise magnitude grows (e.g., from 11.2 to 18.0 at $k = 30$ when

Table 3.4 Number of feature-corrupted data points detected under varying Gaussian noise levels ($\sigma \in \{1.0, 2.0, 5.0\}$)

# Corrupted	$\sigma = 1.0$	$\sigma = 2.0$	$\sigma = 5.0$
10	2.0 ± 1.6	2.4 ± 1.1	2.4 ± 1.1
20	6.2 ± 4.4	9.0 ± 2.6	10.2 ± 2.3
30	11.2 ± 6.0	13.2 ± 3.1	18.0 ± 4.2
40	25.8 ± 2.9	28.4 ± 6.2	30.0 ± 4.7

Table 3.5 Number of detected flipped labels across data modalities

# Flipped	$k = 10$	$k = 20$	$k = 30$	$k = 40$
20 Newsgroups	3.0 ± 1.0	8.8 ± 2.3	16.8 ± 3.4	22.6 ± 2.4
Adult	3.4 ± 2.1	6.8 ± 1.9	11.6 ± 2.4	17.2 ± 3.4

σ increases from 1.0 to 5.0). This sensitivity stems from LiveVal's parameter-space analysis: larger noise produces more significant deviations from the reference trajectory, making corrupted data points more distinguishable through their gradient contributions.

LiveVal also performs better with larger-scale corruptions, identifying 75% of corrupted data at $k = 40$ compared to 24% at $k = 10$ under $\sigma = 5.0$. This behavior arises from LiveVal's adaptive reference mechanism: when more corrupted data points are present, their collective impact creates more distinct parameter trajectories, enabling more reliable detection through reference point comparisons. These properties make LiveVal particularly effective for real-world scenarios where data quality issues often manifest as systematic corruptions affecting multiple data points rather than isolated noise.

Robustness to Different Modalities

Different data modalities present distinct challenges for data valuation due to their inherent structural differences. This experiment evaluates LiveVal's generalizability on text and tabular data.

Experimental Design

We evaluate LiveVal on 20 Newsgroups (text) and Adult (tabular) using a DNN architecture. We flip k labels ($k \in \{10, 20, 30, 40\}$) in each dataset and, following the same protocol as section "Experimental Comparison", evaluate detection performance on a pool of 100 data points (k flipped plus $100 - k$ clean) by counting the number of flipped data points among the k lowest-valued.

Results and Analysis

Table 3.5 reveals modality-dependent detection behavior. For text data (20 Newsgroups), LiveVal's detection rate improves substantially from 30% at $k = 10$ to 56.5% at $k = 40$. This strong scaling can be attributed to the rich feature interactions in text data: corrupted data points tend to create more distinct parameter trajectories due to the high-dimensional, sparse nature of text representations. For tabular data (Adult), LiveVal shows more moderate improvement, from 34% at $k = 10$ to 43% at $k = 40$, reflecting the lower-dimensional and denser feature structure of tabular inputs. Despite these modality-specific variations, LiveVal maintains effective detection performance across both domains, demonstrating its generality as a data valuation framework.

Scalability to Large Models

Modern deep learning often employs large architectures where traditional data valuation methods become computationally impractical. This experiment evaluates LiveVal on Inception V3 (48 layers, approximately 24M parameters) to assess its scalability.

Experimental Design

Following the same label-flipping setup from section "Experimental Comparison", we flip k labels from digit '1' to digit '7' ($k \in \{10, 20, 30, 40\}$) and train Inception V3 on the modified MNIST dataset for one epoch. Detection performance is evaluated on 100 randomly selected data points (k flipped and $100 - k$ clean) by identifying the k lowest-valued.

Results and Analysis

Table 3.6 shows LiveVal's detection performance on Inception V3. With increasing corruption levels, LiveVal maintains stable detection rates, reaching 56% for $k = 40$ even with just one epoch of training. This early detection capability is essential for identifying harmful data before they substantially degrade model quality.

Table 3.6 LiveVal's detection performance on Inception V3 (24M parameters)

# Flipped	$k = 10$	$k = 20$	$k = 30$	$k = 40$
# Detected	3.0 ± 2.2	8.6 ± 4.2	11.4 ± 5.9	22.6 ± 2.4
Detection rate (%)	30	43	38	56

LiveVal also demonstrates significant computational advantages at this scale. It completes the evaluation in 7 min 32 s on a single NVIDIA A40 GPU, which is remarkable given Inception V3's 24M parameters, compared to LOO's estimated runtime of hundreds of hours. LiveVal requires only 31.18 GB of memory, whereas influence functions would need approximately 2.3 TB for Hessian computation. These efficiency gains confirm that LiveVal remains practical for large-scale architectures where traditional methods become prohibitive.

The robustness experiments confirm that LiveVal generalizes effectively across feature-level corruption, multiple data modalities, and large model architectures. The next subsection examines the sensitivity of LiveVal's adaptive window mechanism to hyperparameter choices.

Hyperparameter Sensitivity Analysis

The adaptive window mechanism is central to LiveVal's real-time performance, and the thresholds $\varepsilon_{\min}$ and $\varepsilon_{\max}$ directly control when the algorithm adjusts its temporal scope. This subsection evaluates how sensitive detection performance is to these hyperparameter choices and visualizes the adaptive window dynamics during training.

Experimental Design

Using the same experimental setup as section "Experimental Comparison", we conduct sensitivity analysis on $\varepsilon_{\min}$ and $\varepsilon_{\max}$ under the scenario of 40 flipped labels. We vary $\varepsilon_{\min} \in \{0.0001, 0.00032, 0.001, 0.0032, 0.01\}$ and $\varepsilon_{\max} \in \{0.01, 0.032, 0.1, 0.32, 1.0\}$, covering three orders of magnitude to capture both conservative and aggressive adaptation behaviors. Other parameters are held fixed: learning rate 0.1, $\delta_{\min} = 1$, $\delta_{\max} = 100$, $\Delta\delta = 1$. Each configuration is evaluated by the detection rate, defined as the fraction of flipped labels correctly identified among the k lowest-valued data.

Results and Analysis

Figure 3.4 reveals several insights about the hyperparameter landscape. The peak detection rate (0.86) occurs at $\varepsilon_{\min} = 0.0032$ and $\varepsilon_{\max} = 0.32$, a ratio of approximately 1:100 that reflects a clear separation between stable and volatile training stages. More importantly, LiveVal achieves robust performance across a broad range of middle-range configurations, with detection rates consistently above 0.70, demonstrating that the method does not require precise hyperparameter tuning in practice.

Performance degrades predictably at parameter extremes. When $\varepsilon_{\min} \leq 0.0001$, the algorithm becomes overly conservative, adapting windows too slowly and failing to shrink them during convergence phases; detection rates drop to the range of 0.05 to 0.09 because the method misses fine-grained harmful data patterns. When

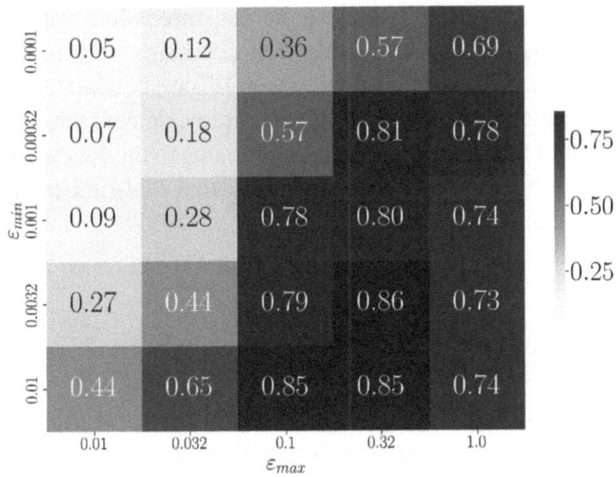

Fig. 3.4 LiveVal detection rate across different $(\varepsilon_{\min}, \varepsilon_{\max})$ configurations. The peak performance (0.86) occurs at $\varepsilon_{\min} = 0.0032$, $\varepsilon_{\max} = 0.32$

$\varepsilon_{\max} \geq 1.0$, the algorithm treats normal training fluctuations as volatility, triggering unnecessary window expansion that degrades the reference point quality, and performance plateaus around 0.70.

Conclusion

Real-time data influence measurement is the first pillar of the theory of data-level unlearning, and this chapter has established its theoretical and algorithmic foundations. The trajectory-based valuation framework introduced in section "Trajectory-Based Data Valuation Framework" moves beyond traditional post-hoc methods by modeling how each data point steers the optimization trajectory as training unfolds. The LiveVal algorithm (section "Real-Time Measurement with Adaptive Reference Points") realizes this framework in practice through an adaptive reference point mechanism and a dual-queue architecture (section "Real-Time Implementation via Dual Queues") that keeps both storage and computation within practical bounds.

Three theoretical properties underpin the reliability of LiveVal's influence estimates. Directional alignment and value boundedness (Theorem 3.1) ensure that valuations are semantically meaningful and scale-invariant. Stability under stochastic training (Theorem 3.2) guarantees that volatility shrinks with the learning rate, producing increasingly reliable estimates as training progresses. Temporal fairness (Theorem 3.3) eliminates the systematic bias that causes unnormalized gradient-based methods to overvalue early-training data, ensuring that data points are assessed by their actual optimization contributions rather than their temporal position. These properties provide a principled basis for identifying and quantifying the influence of harmful or low-quality data points during training.

Experimental evaluation across three data modalities, multiple model architectures, and diverse corruption scenarios confirms the practical effectiveness of LiveVal. The method achieves a 180× speedup over Leave-One-Out while maintaining competitive detection accuracy, detects the majority of corrupted data within the first few training epochs, and scales to models with 24M parameters on commodity hardware. These results validate that real-time data valuation is both theoretically sound and practically viable.

References

1. Song, T., Tong, Y., Wei, S.: Profit allocation for federated learning. In: IEEE International Conference on Big Data (Big Data), pp. 2577–2586. IEEE (2019)
2. Sim, R.H.L., et al.: Collaborative machine learning with incentive-aware model rewards. In: International Conference on Machine Learning. PMLR, pp. 8927–8936 (2020)
3. Ghorbani, A., Zou, J.: Data Shapley: equitable valuation of data for machine learning. In: International Conference on Machine Learning. PMLR, pp. 2242–2251 (2019)
4. Kwon, Y., Zou, J.: Beta Shapley: a unified and noise-reduced data valuation framework for machine learning. In: International Conference on Artificial Intelligence and Statistics. PMLR, pp. 8780–8802 (2022)
5. Jia, R., et al.: Towards efficient data valuation based on the Shapley value. In: International Conference on Artificial Intelligence and Statistics. PMLR, pp. 1167–1176 (2019)
6. Koh, P.W., Liang, P.: Understanding black-box predictions via influence functions. In: International Conference on Machine Learning. PMLR, pp. 1885–1894 (2017)
7. Choe, S.K., et al.: What is your data worth to GPT? LLM-scale data valuation with influence functions. arXiv preprint arXiv:2405.13954 (2024)
8. Nguyen, T.D., et al.: FLAME: Taming backdoors in federated learning. In: 31st USENIX Security Symposium (USENIX Security 22), pp. 1415–1432 (2022)
9. Koh, P.W., Steinhardt, J., Liang, P.: Stronger data poisoning attacks break data sanitization defenses. In: Machine Learning, pp. 1–47 (2022)
10. Goodfellow, I.J., Shlens, J., Szegedy, C.: Explaining and harnessing adversarial examples. arXiv preprint arXiv:1412.6572 (2014)
11. Zhang, C., et al.: Understanding deep learning requires rethinking generalization. arXiv preprint arXiv:1611.03530 (2016)
12. Cubuk, E.D., et al.: Autoaugment: learning augmentation policies from data. arXiv preprint arXiv:1805.09501 (2018)
13. Koh, P.W.W., et al.: On the accuracy of influence functions for measuring group effects. In: Advances in Neural Information Processing Systems, vol. 32 (2019)
14. Pruthi, G., et al.: Estimating training data influence by tracing gradient descent. In: Advances in Neural Information Processing Systems, vol. 33, pp. 19920–19930 (2020)
15. Wang, J.T., Jia, R.: Data Banzhaf: a robust data valuation framework for machine learning. In: International Conference on Artificial Intelligence and Statistics. PMLR, pp. 6388–6421 (2023)
16. Paul, M., Ganguli, S., Dziugaite, G.K.: Deep learning on a data diet: finding important examples early in training. In: Advances in Neural Information Processing Systems, vol. 34, pp. 20596–20607 (2021)

17. Pascanu, R., Mikolov, T., Bengio, Y.: On the difficulty of training recurrent neural networks. In: International Conference on Machine Learning. PMLR, pp. 1310–1318 (2013)
18. Dua, D., Graff, C.: UCI Machine Learning Repository. http://archive.ics.uci.edu/ml (2019)
19. Lang, K.: Newsweeder: learning to filter netnews. In: International Conference on Machine Learning, pp. 331–339 (1995)
20. LeCun, Y., Cortes, C., Burges, C., et al.: MNIST handwritten digit database. http://yann.lecun.com/exdb/mnist (2010)

Chapter 4
Time-Varying Data Influence Measurement

Abstract Existing data influence methods summarize each training data point's contribution as a single global score evaluated at the final model state. This global view indicates whether a data point matters, but not when during training it exerted its effect, forcing unlearning methods to treat the entire training trajectory as affected and limiting the interpretability of data–model interactions. This chapter proposes Time-Varying Influence Measurement (TVIM), a framework that quantifies the accumulated influence of a training data point within arbitrary sub-intervals of the training process. TVIM operates in two phases: during training, it records essential trajectory information, including parameter states and update directions; after training, it analytically replays this information to reconstruct how a data point shaped the model over any chosen window, without retraining. By localizing influence to specific training stages, TVIM provides two advantages over global methods. First, it enables more efficient unlearning by restricting corrections to the minimal affected window rather than the entire trajectory. Second, it produces temporal influence profiles that reveal how and when a data point interacted with the optimization dynamics, improving interpretability. We derive estimation error guarantees under general non-convex conditions and demonstrate that the approach identifies critical influence windows useful for targeted data removal and model auditing.

Motivation and Contribution

Understanding how individual training data points influence a trained model is central to machine unlearning, data debugging, and model auditing [1–3]. The prevailing tools for this purpose, including influence functions [1], Shapley values [2, 4], and their variants, all adopt a *global* perspective: they summarize each data point's contribution as a single scalar evaluated at the final model state. This global view compresses the entire history of a data point's interaction with the optimization process into one aggregate number. While useful for ranking data by overall importance, it fundamentally limits both the efficiency and the interpretability of downstream tasks that depend on influence estimates.

J. Xu and X. Jia, *Machine Unlearning*, Wireless Networks,
https://doi.org/10.1007/978-3-032-30978-5_4

The efficiency limitation is most apparent in machine unlearning. When a data point is identified as malicious, corrupted, or otherwise undesirable, global influence scores indicate *whether* the data point matters but not *when* during training it exerted its effect [5–7]. Without temporal localization, the only safe unlearning strategy is to treat the entire training trajectory as contaminated: either retrain from scratch or apply a single correction at the final state that may be inaccurate and inefficient because the influence was not uniformly distributed [8–10]. If, instead, we could determine that a data point's influence was concentrated in a specific window, say steps t_1 through t_2, then unlearning could be restricted to that window alone. Only the parameter updates within $[t_1, t_2]$ would need to be revised while the rest of the trajectory is preserved, making the removal both cheaper and more precise.

The interpretability limitation is equally significant. A single global score provides no insight into the mechanism through which a data point shaped the model. It cannot answer questions such as: Did a mislabeled data point cause damage primarily during early learning when decision boundaries were forming, or during late fine-tuning when the model memorized it? Did a poisoned data point exert steady influence throughout training, or did its effect spike within a narrow window? Answering such questions requires decomposing influence along the temporal axis, producing a profile that reveals *how* and *when* a data point interacted with the optimization dynamics rather than merely *how much* it contributed in aggregate.

To address both limitations, we propose Time-Varying Influence Measurement (TVIM), a framework for quantifying the accumulated influence of a training data point within arbitrary sub-intervals of the training process. TVIM operates in two phases. During training, the system records essential trajectory information, including parameter states and update directions, without altering the learning process. When an influence query arises after training, this stored information is analytically replayed to estimate how excluding a specific data point would have altered the model's parameters over a chosen window [10, 11]. In this way, TVIM approximates a counterfactual learning scenario without retraining from scratch.

The contributions of this chapter are threefold. First, we establish a window-level formulation of data influence that generalizes global and instantaneous notions into cumulative, temporally bounded quantities, providing both a more efficient basis for targeted unlearning and a more interpretable characterization of interactions between data and the model. Second, we develop the two-phase TVIM framework that records trajectory information once and enables retrospective computation of data influence over arbitrary training windows, so that unlearning can be localized to the minimal affected interval rather than applied to the entire trajectory. Third, we present error bounds and analytical guarantees showing that time-varying influence can be approximated with controlled accuracy even in non-convex settings typical of deep models; these results guide practical choices for the correction window size and quantify the trade-off between estimation precision and computational cost [12].

Overall, this chapter extends data influence analysis from a global, single-score paradigm to a temporally resolved framework, enabling both more efficient unlearning through window-localized corrections and deeper interpretability of how training data shape model behavior across the optimization trajectory.

System Models and Problem Statement

Formalizing time-varying influence requires two ingredients: a precise description of the training dynamics and a mathematical definition of what it means for a data point to influence a specific training window. This section provides both. Section "System Model" establishes the system model, defining the SGD training process, the counterfactual process that excludes a single data point, and the existing notions of data influence that our framework generalizes. Section "Problem Statement" then formalizes the window-level influence quantity that TVIM aims to estimate.

System Model

Let $\mathcal{Z} = \mathcal{X} \times \mathcal{Y}$ denote the space of observations, where $\mathcal{X} \subseteq \mathbb{R}^d$ is the input space and $\mathcal{Y}$ is the output space. Given a training set $\mathcal{D} = \{z_i\}_{i=1}^N$ of i.i.d. observations $z_i = (x_i, y_i) \in \mathcal{Z}$, a model $f : \mathcal{X} \times \Theta \to \mathcal{Y}$ parameterized by $\theta \in \Theta \subseteq \mathbb{R}^p$, and a loss function $\ell : \mathcal{Z} \times \Theta \to \mathbb{R}$, we formulate the learning problem as empirical risk minimization [13]:

$$\hat{\theta} = \arg\min_{\theta \in \Theta} \frac{1}{N} \sum_{i=1}^{N} \ell(z_i; \theta). \tag{4.1}$$

Definition 4.1 (*Stochastic Gradient Descent (SGD)*) Let $g(z; \theta) = \nabla_\theta \ell(z; \theta)$, and let SGD start from $\theta^{[0]}$. The update rule for mini-batch SGD at step t is [14, 15]:

$$\theta^{[t+1]} = \theta^{[t]} - \frac{\eta_t}{|S_t|} \sum_{i \in S_t} g(z_i; \theta^{[t]}), \quad 0 \leq t \leq T - 1, \tag{4.2}$$

where $S_t \subseteq \{1, \ldots, N\}$ represents the mini-batch of indices at step t, η_t is the learning rate at step t, and T denotes the total number of SGD steps.

Definition 4.2 (*Influence Function* [1]) The influence function measures the impact of removing a single training data point z_j on the optimal model parameters $\hat{\theta}$. It is defined as $\hat{\theta}_{-j} - \hat{\theta}$, where $\hat{\theta}_{-j} = \arg\min_\theta \sum_{i=1, i \neq j}^{N} \ell(z_i; \theta)$. For strongly convex loss functions, it can be approximated as:

$$\hat{\theta}_{-j} - \hat{\theta} \approx -\hat{H}^{-1} \nabla_\theta \ell(z_j; \hat{\theta}), \tag{4.3}$$

where $\hat{H} = \frac{1}{N} \sum_{z \in \mathcal{D}} \nabla^2 \ell(z; \hat{\theta})$ is the Hessian of the loss at the optimal parameters.

Definition 4.3 (*Counterfactual SGD*) The counterfactual SGD process excludes the j-th training data point in order to quantify its influence on the learning process [10, 11]. Starting from $\theta_{-j}^{[0]} = \theta^{[0]}$, the parameters are updated at each step t using:

$$\theta_{-j}^{[t+1]} = \theta_{-j}^{[t]} - \frac{\eta_t}{|S_t|} \sum_{i \in S_t \setminus \{j\}} g(z_i; \theta_{-j}^{[t]}), \quad 0 \leq t \leq T - 1. \tag{4.4}$$

Definition 4.4 (*SGD-Influence* [11]) The SGD-influence of training data point $z_j \in \mathcal{D}$ within t steps is defined as $\theta_{-j}^{[t]} - \theta^{[t]}$.

While the influence function [1] provides insights at the optimum, SGD-Influence [11] measures the impact of excluding a specific training data point z_j throughout the SGD training process. Neither formulation, however, quantifies influence within an arbitrary sub-interval of training. The next subsection formalizes this window-level problem.

Problem Statement

Our goal is to estimate the impact of training data points during an arbitrary time window $[t_1, t_2]$ within the overall training process $[0, T]$, where $0 \leq t_1 < t_2 \leq T$. We formalize this goal with a counterfactual question: how would the model's parameters change during the interval $[t_1, t_2]$ if a specific data point z_j were not used?

Definition 4.5 (*Parameter Change in Time Window*) For a time window $[t_1, t_2]$ during SGD training, the parameter change estimates the contribution of a training data point z_j as:

$$\Delta\theta_{-j}^{[t_1,t_2]} = \left(\theta_{-j}^{[t_2]} - \theta_{-j}^{[t_1]}\right) - \left(\theta^{[t_2]} - \theta^{[t_1]}\right), \tag{4.5}$$

where $(\theta^{[t_2]} - \theta^{[t_1]})$ represents the parameter changes under standard SGD within $[t_1, t_2]$, and $(\theta_{-j}^{[t_2]} - \theta_{-j}^{[t_1]})$ represents the parameter changes over the same interval when excluding data point z_j.

For the special case $[0, t]$, starting from the beginning of training, this simplifies to:

$$\Delta\theta_{-j}^{[0,t]} = \left(\theta_{-j}^{[t]} - \theta_{-j}^{[0]}\right) - \left(\theta^{[t]} - \theta^{[0]}\right) = \theta_{-j}^{[t]} - \theta^{[t]}. \tag{4.6}$$

For brevity, we denote $\Delta\theta_{-j}^{[0,t]} = \Delta\theta_{-j}^{[t]}$.

The next section develops a closed-form approximation of this quantity and analyzes its estimation error.

Design of Time-Varying Influence Measurement

Estimating the time-varying influence $\Delta\theta_{-j}^{[t_1,t_2]}$ defined in the previous section requires approximating a counterfactual training trajectory without actually retraining the model, a central challenge shared by influence-function and trajectory-tracing approaches [1, 3, 10, 11]. This section develops the estimation in two stages. Section "Influence on Model Parameters" derives a closed-form approximation of the parameter-level influence by recursively expanding the difference between the standard and leave-one-out SGD trajectories, and establishes an error bound for this approximation under non-convex conditions. Section "Influence on Model Behaviors" then extends the framework from raw parameter changes to observable model behaviors (loss, predictions, feature importance) through a query-based projection mechanism.

Influence on Model Parameters

We aim to estimate the parameter change due to the absence of data point z_j over the time window $[t_1, t_2]$, where $0 \leq t_1 < t_2 \leq T$:

$$\Delta\theta_{-j}^{[t_1,t_2]} = \left(\theta_{-j}^{[t_2]} - \theta_{-j}^{[t_1]}\right) - \left(\theta^{[t_2]} - \theta^{[t_1]}\right) = \left(\theta_{-j}^{[t_2]} - \theta^{[t_2]}\right) - \left(\theta_{-j}^{[t_1]} - \theta^{[t_1]}\right). \quad (4.7)$$

Consider the normal SGD update for step t $(0 \leq t \leq T-1)$, including all data points:

$$\theta^{[t+1]} = \theta^{[t]} - \frac{\eta_t}{|S_t|} \sum_{i \in S_t} g(z_i;\, \theta^{[t]}). \quad (4.8)$$

Consider the SGD update excluding data point z_j:

$$\theta_{-j}^{[t+1]} = \theta_{-j}^{[t]} - \frac{\eta_t}{|S_t|} \sum_{i \in S_t \setminus \{j\}} g(z_i;\, \theta_{-j}^{[t]}). \quad (4.9)$$

Calculate the difference between the two updates:

$$\theta_{-j}^{[t+1]} - \theta^{[t+1]} = \left(\theta_{-j}^{[t]} - \theta^{[t]}\right) - \frac{\eta_t}{|S_t|}\Big(\sum_{i \in S_t \setminus \{j\}} g(z_i;\, \theta_{-j}^{[t]}) - \sum_{i \in S_t} g(z_i;\, \theta^{[t]})\Big). \quad (4.10)$$

Approximate the gradient differences using a first-order Taylor expansion:

$$g(z_i;\, \theta_{-j}^{[t]}) - g(z_i;\, \theta^{[t]}) \approx \nabla_\theta g(z_i;\, \theta^{[t]})^T \left(\theta_{-j}^{[t]} - \theta^{[t]}\right), \quad (4.11)$$

where $\nabla_\theta g(z_i;\, \theta^{[t]})$ is the gradient of $g(z_i;\, \theta)$ with respect to θ, evaluated at $\theta^{[t]}$. Define the approximate Hessian matrix $H^{[t]}$ as the average of the outer products of these gradients over the mini-batch:

$$H^{[t]} = \frac{1}{|S_t|} \sum_{i \in S_t} \nabla_\theta g(z_i;\, \theta^{[t]})^T. \quad (4.12)$$

Using this definition in Eqs. (4.12) and (4.11), we have:

$$\frac{1}{|S_t|}\sum_{i\in S_t}\left(g(z_i;\,\theta_{-j}^{[t]}) - g(z_i;\,\theta^{[t]})\right) \approx H^{[t]}\left(\theta_{-j}^{[t]} - \theta^{[t]}\right). \tag{4.13}$$

Substituting this approximation into Eq. (4.10), we have:

$$\theta_{-j}^{[t+1]} - \theta^{[t+1]} \approx (I - \eta_t H^{[t]})\left(\theta_{-j}^{[t]} - \theta^{[t]}\right) + \tilde{\mathbf{1}}_j^{[t]}, \tag{4.14}$$

where $\tilde{\mathbf{1}}_j^{[t]} = \mathbf{1}_{j\in S_t}\frac{\eta_t}{|S_t|}g(z_j;\,\theta^{[t]})$ and $\mathbf{1}_{j\in S_t}$ is an indicator function that equals 1 if $j \in S_t$ and 0 otherwise. Let $Z_t = I - \eta_t H^{[t]}$, and recursively apply this relation over the interval $[t_1, t_2]$:

$$\theta_{-j}^{[t_2]} - \theta^{[t_2]} \approx Z_{t_2-1}Z_{t_2-2}\cdots Z_{t_1}\left(\theta_{-j}^{[t_1]} - \theta^{[t_1]}\right) + \sum_{t=t_1}^{t_2-1} Z_{t_2-1}Z_{t_2-2}\cdots Z_{t+1}\,\tilde{\mathbf{1}}_j^{[t]}. \tag{4.15}$$

Combining Eqs. (4.7) and (4.15), we obtain:

$$\Delta\theta_{-j}^{[t_1,t_2]} \approx \left(\prod_{k=t_1}^{t_2-1} Z_k - I\right)\left(\theta_{-j}^{[t_1]} - \theta^{[t_1]}\right) + \sum_{t=t_1}^{t_2-1}\left(\prod_{k=t+1}^{t_2-1} Z_k\right)\tilde{\mathbf{1}}_j^{[t]}. \tag{4.16}$$

We use Eq. (4.16) for the interval $[0, t_1]$ with $\theta_{-j}^{[0]} = \theta^{[0]}$ to get $(\theta_{-j}^{[t_1]} - \theta^{[t_1]})$:

$$\Delta\theta_{-j}^{[0,t_1]} = \theta_{-j}^{[t_1]} - \theta^{[t_1]} \approx \sum_{t=0}^{t_1-1}\left(\prod_{k=t+1}^{t_1-1} Z_k\right)\tilde{\mathbf{1}}_j^{[t]}. \tag{4.17}$$

Substituting Eq. (4.17) back into Eq. (4.16), we obtain:

$$\Delta\theta_{-j}^{[t_1,t_2]} \approx \left(\prod_{k=t_1}^{t_2-1} Z_k - I\right)\left(\sum_{t=0}^{t_1-1}\left(\prod_{k=t+1}^{t_1-1} Z_k\right)\tilde{\mathbf{1}}_j^{[t]}\right) + \sum_{t=t_1}^{t_2-1}\left(\prod_{k=t+1}^{t_2-1} Z_k\right)\tilde{\mathbf{1}}_j^{[t]}. \tag{4.18}$$

We define the estimated parameter change as:

$$\widehat{\Delta\theta}_{-j}^{[t_1,t_2]} = \left(\prod_{k=t_1}^{t_2-1} Z_k - I\right)\left(\sum_{t=0}^{t_1-1}\left(\prod_{k=t+1}^{t_1-1} Z_k\right)\tilde{\mathbf{1}}_j^{[t]}\right) + \sum_{t=t_1}^{t_2-1}\left(\prod_{k=t+1}^{t_2-1} Z_k\right)\tilde{\mathbf{1}}_j^{[t]}. \tag{4.19}$$

Estimation Error Analysis Without Convexity Assumptions

The first-order Taylor approximation used in the recursive expansion introduces error that accumulates over the training window. Under standard non-convex optimization assumptions, the following bound characterizes this accumulation:

$$\mathbb{E}\Big[\Big\| \Delta\theta_{-j}^{[t_1,t_2]} - \widehat{\Delta\theta}_{-j}^{[t_1,t_2]} \Big\|\Big] \leq \frac{\tilde{B}}{M_H} \left(e^{M_H \eta_{\max}(t_2+1)} + e^{M_H \eta_{\max}(t_1+1)} - 2 \right), \tag{4.20}$$

where M_H is the upper bound on the norm of the Hessian matrix $H^{[t]}$, $\eta_{\max}$ is the maximum learning rate, and $\tilde{B} = \frac{L_H M^2}{2} + \epsilon_H M$ encapsulates constants related to the Hessian's Lipschitz continuity and approximation error.

This bound shows that while the estimation error may grow exponentially with t_1 and t_2, it can be controlled under appropriate conditions. This result provides theoretical support for the reliability of TVIM without assuming convexity. For detailed derivations and assumptions, see section "Estimation Error Analysis".

The derivation above shows that the time-varying parameter influence can be expressed as a recursive product of local update matrices, and its estimation error is bounded even in non-convex settings. However, raw parameter-space influence is not always the most informative quantity. The next subsection extends the framework to measure how data-point exclusion affects observable model behaviors such as loss values, predictions, and feature importance.

Influence on Model Behaviors

The parameter-level analysis in section "Influence on Model Parameters" reveals how data points shift the model's internal state, yet practitioners often care more about externally observable quantities such as test loss, predictions, or feature attributions. To bridge this gap, TVIM provides a query-based formulation that projects parameter-space influence onto user-specified directions, enabling flexible and interpretable measurement of how individual data points affect various aspects of model performance.

Definition 4.6 (*Query-based Influence Measurement*) Let $q\colon [0, T] \to \mathbb{R}^p$ be a query function that maps time t to a query vector $q(t) \in \mathbb{R}^p$. The query-based influence measurement for a training data point z_j over the time window $[t_1, t_2]$ is defined as:

$$Q_{-j}^{[t_1,t_2]}(q) = \langle q(t_2),\ \Delta\theta_{-j}^{[t_2]} \rangle - \langle q(t_1),\ \Delta\theta_{-j}^{[t_1]} \rangle, \tag{4.21}$$

where $\Delta\theta_{-j}^{[t]} = \theta_{-j}^{[t]} - \theta^{[t]}$ represents the parameter change at time t and $\langle \cdot, \cdot \rangle$ denotes the standard inner product in $\mathbb{R}^p$.

This measure estimates the influence of data point z_j on the model's behavior as projected onto the query directions. For example, by setting $q(t) = \nabla_\theta \ell(z_{\text{test}};\ \theta^{[t]})$, we can measure the impact of a training data point on the model's loss for a test data point z_{test}:

$$\begin{aligned} Q_{-j}^{[t_1,t_2]}(q) &= \left\langle \nabla_\theta \ell(z_{\text{test}};\ \theta^{[t_2]}),\ \Delta\theta_{-j}^{[t_2]} \right\rangle - \left\langle \nabla_\theta \ell(z_{\text{test}};\ \theta^{[t_1]}),\ \Delta\theta_{-j}^{[t_1]} \right\rangle \\ &\approx \left[\ell(z_{\text{test}};\ \theta_{-j}^{[t_2]}) - \ell(z_{\text{test}};\ \theta_{-j}^{[t_1]})\right] - \left[\ell(z_{\text{test}};\ \theta^{[t_2]}) - \ell(z_{\text{test}};\ \theta^{[t_1]})\right]. \end{aligned} \tag{4.22}$$

Different choices of q enable analysis of various model characteristics. Setting $q = \nabla_\theta f(x_{\text{test}};\ \theta^{[t]})$ measures prediction changes, $q = e_i$ (standard basis vector) examines individual parameter importance, and $q = \nabla_\theta \ell(z_j;\ \theta^{[t]})$ assesses gradient alignments.

Implementations

The TVIM framework is realized through two algorithmic phases: a training phase that records trajectory information alongside standard SGD, and an inference phase that retrospectively computes influence for any data point over any time window. This section describes both phases in detail. Section "Standard Implementation" presents the standard implementation that stores per-step information, and section "Checkpoint-Based Implementation of TVIM" introduces a checkpoint-based variant that reduces memory requirements by storing information only at selected intervals.

Standard Implementation

The standard implementation stores trajectory information at every training step, providing full temporal resolution for retrospective influence queries. It consists of a training phase that augments standard SGD with lightweight logging, and an inference phase that reconstructs influence by propagating query vectors backward through the stored trajectory.

Training Phase of TVIM

The training phase of TVIM records the information needed for retrospective influence computation without modifying the SGD process itself. At each step, the algorithm stores the mini-batch indices, the learning rate, and the updated parameters, producing an information sequence that fully characterizes the training trajectory. Algorithm 2 details this process.

Algorithm 2 Training Phase of TVIM

Require: Training dataset $\mathcal{D} = \{z_n\}_{n=1}^{N}$, learning rate η_t, batch size $|S_t|$, training steps T
Ensure: Stored information A

Initialize model parameters $\theta^{[0]}$
Initialize an empty sequence A
for $t = 1$ to T **do**
  $S_t = \text{SampleBatch}(\mathcal{D}, |S_t|)$
  $\theta^{[t+1]} = \theta^{[t]} - \frac{\eta_t}{|S_t|} \sum_{i \in S_t} g(z_i; \theta^{[t]})$
  $A[t] = \{S_t, \eta_t, \theta^{[t+1]}\}$
end for
return A

The time complexity of this algorithm is $O(T \cdot |S_t| \cdot p)$, where p is the number of model parameters. The space complexity is $O\big(T(|S_t| + p)\big)$. Periodic data compression can be performed during training to balance memory efficiency with information utilization for later influence analysis.

Inference Phase of TVIM

The inference phase uses the stored trajectory information to compute the influence of a specific data point on a query across an arbitrary time window. Algorithm 3 uses two key variables, $u_2^{[t]}$ and $u_1^{[t]}$, which propagate $q(t_2)$ and $q(t_1)$ backwards through time while incorporating the Z_k matrices. The algorithm computes Q by summing the inner products of $(u_2^{[t]} - u_1^{[t]})$ with $\tilde{\mathbf{1}}_j^{[t]}$ at each time step. When expanded, this sum matches the structure of $Q_{-j}^{[t_1,t_2]}(q)$ as defined earlier, with the accumulated terms corresponding to the contributions from $\Delta\theta_{-j}^{[t_2]}$ and $\Delta\theta_{-j}^{[t_1]}$. This approach efficiently computes the influence without explicitly performing large matrix multiplications.

Algorithm 3 Inference Phase of TVIM

Require: Stored information A, query function q, time window $[t_1, t_2]$, specified data point z_j
Ensure: Estimated influence Q for data point z_j

Initialize $Q \leftarrow 0$, $u_1^{[t_2-1]} \leftarrow 0$
Initialize $u_2^{[t_2-1]} \leftarrow q(t_2)$
for $t = t_2 - 1$ **downto** 0 **do**
  if $j \in S_t$ **then**
    $Q \leftarrow Q + \left\langle (u_2^{[t]} - u_1^{[t]}), \frac{\eta_t}{|S_t|} g(z_j; \theta^{[t]}) \right\rangle$
  end if
  $u_1^{[t-1]} \leftarrow u_1^{[t]} - \eta_t H^{[t]} u_1^{[t]}$
  $u_2^{[t-1]} \leftarrow u_2^{[t]} - \eta_t H^{[t]} u_2^{[t]}$
  if $t = t_1$ **then**
    $u_1^{[t-1]} \leftarrow q(t_1)$
  end if
end for
return Q

The time complexity of Algorithm 3 is $O(T|S_t|p)$ and the space complexity is $O\big(T(|S_t| + p)\big)$, where T is the number of training steps. TVIM avoids the computationally intensive direct computation and storage of the Hessian matrix, which typically requires $O(Tp^2)$ operations. Instead, it computes Hessian-vector products $H^{[t]}u = \nabla_\theta \langle u, g(z; \theta^{[t]}) \rangle$ in $O(|S_t|p)$ per step, making it practical for modern large-scale models.

Checkpoint-Based Implementation of TVIM

Storing parameters at every training step may be prohibitive for large models or long training runs. To balance storage overhead and computational efficiency, we propose a checkpoint-based implementation of TVIM that significantly reduces storage requirements while maintaining the ability to compute accurate influence values for any time window.

Instead of storing parameters at every training step, we store checkpoints at regular intervals (e.g., epoch boundaries) along with essential training metadata (batch indices and learning rates). When computing influence for a time window $[t_1, t_2]$, we efficiently recover necessary parameters by loading the nearest checkpoint before t_1, reconstructing the parameter trajectory up to t_2, and storing intermediate parameters required for influence computation. The checkpoint interval provides a configurable trade-off between storage overhead and computational cost. More frequent checkpoints reduce recomputation but increase storage, while fewer checkpoints save storage at the cost of more recomputation.

Algorithm 4 Training with Checkpoints

Require: Training dataset $\mathcal{D} = \{z_n\}_{n=1}^N$, learning rate η_t, batch size $|S_t|$, training steps T, checkpoint interval C
Ensure: Stored checkpoints and metadata M
Initialize model parameters $\theta^{[0]}$
Initialize metadata storage $M \leftarrow \{\}$ {Store checkpoints, batch indices, learning rates}
for $t = 0$ **to** $T - 1$ **do**
  $S_t \leftarrow \text{DataBatch}(\mathcal{D}, |S_t|)$
  $\theta^{[t+1]} \leftarrow \theta^{[t]} - \frac{\eta_t}{|S_t|} \sum_{i \in S_t} g(z_i; \theta^{[t]})$
  $M.indices[t] \leftarrow S_t$ {Store batch indices}
  $M.lr[t] \leftarrow \eta_t$ {Store learning rate}
  if $t \bmod C = 0$ **or** $t = T - 1$ **then**
    $M.checkpoints[t] \leftarrow \theta^{[t+1]}$ {Store checkpoint}
  end if
end for
return M

Algorithm 5 TVIM Data Influence with Checkpoints

Require: Metadata M, query function q, time window $[t_1, t_2]$, data z_j
Ensure: Estimated influence Q
$Q \leftarrow 0, u_1^{[t_2-1]} \leftarrow 0, u_2^{[t_2-1]} \leftarrow q(t_2)$
$c_1 \leftarrow \max\{t : t \le t_1 \text{ and } t \in M.checkpoints\}$ {Find nearest checkpoint before t_1}
$\theta^{[c_1]} \leftarrow M.checkpoints[c_1]$
{Compute and store all necessary parameters from checkpoint to t_2}
Initialize parameter storage $P \leftarrow \{\}$
for $t = c_1$ **to** $t_2 - 1$ **do**
  $S_t \leftarrow M.indices[t]$
  $\eta_t \leftarrow M.lr[t]$
  if $t \in M.checkpoints$ **then**
    $\theta^{[t]} \leftarrow M.checkpoints[t]$
  end if
  $\theta^{[t+1]} \leftarrow \theta^{[t]} - \frac{\eta_t}{|S_t|} \sum_{i \in S_t} g(z_i; \theta^{[t]})$
  $P[t] \leftarrow \theta^{[t]}$ {Store parameter for influence computation}
end for
for $t = t_2 - 1$ **downto** t_1 **do**
  if $j \in M.indices[t]$ **then**
    $Q \leftarrow Q + \left\langle (u_2^{[t]} - u_1^{[t]}), \frac{M.lr[t]}{|M.indices[t]|} g(z_j; P[t]) \right\rangle$
  end if
  $H^{[t]} \leftarrow \frac{1}{|M.indices[t]|} \sum_{i \in M.indices[t]} \nabla_\theta g(z_i; P[t])$
  $u_1^{[t-1]} \leftarrow u_1^{[t]} - M.lr[t] H^{[t]} u_1^{[t]}$
  $u_2^{[t-1]} \leftarrow u_2^{[t]} - M.lr[t] H^{[t]} u_2^{[t]}$
  if $t = t_1$ **then**
    $u_1^{[t-1]} \leftarrow q(t_1)$
  end if
end for
return Q

Theoretical Analysis

The TVIM estimator introduced in the preceding sections relies on first-order Taylor approximations that accumulate error over the training trajectory. Quantifying this error is essential for determining when the estimated influence is reliable enough to guide unlearning decisions. Section "Estimation Error Analysis" presents the full proof of the estimation error bound previewed in section"Estimation Error Analysis Without Convexity Assumptions", establishing that time-varying influence can be approximated with controlled accuracy under standard non-convex assumptions. Section "Time-Varying Influence Toolkit" then develops a toolkit of query vectors that instantiate TVIM for specific aspects of model behavior, including loss values, prediction changes, feature importance, and parameter importance.

Estimation Error Analysis

Theorem 4.1 (Error Bound for TVIM Parameter Change) *Let $\Delta\theta_{-j}^{[t_1,t_2]}$ be the true influence of excluding data point z_j on the model parameters over the interval $[t_1, t_2]$ during SGD training. Let $\widehat{\Delta\theta}_{-j}^{[t_1,t_2]}$ be its approximation using TVIM. Under the following assumptions:*

(A1) Lipschitz Continuity of Gradient: The gradient $\nabla\ell(z_i;\theta)$ is Lipschitz continuous with constant L_g: $\|\nabla\ell(z_i;\theta_1) - \nabla\ell(z_i;\theta_2)\| \leq L_g\|\theta_1 - \theta_2\|$, $\forall\theta_1, \theta_2 \in \Theta, \forall i$.
(A2) Lipschitz Continuity of Hessian: The Hessian $\nabla^2\ell(z_i;\theta)$ is Lipschitz continuous with constant L_H: $\|\nabla^2\ell(z_i;\theta_1) - \nabla^2\ell(z_i;\theta_2)\| \leq L_H\|\theta_1 - \theta_2\|$, $\forall\theta_1, \theta_2 \in \Theta, \forall i$.
(A3) Learning Rate Bound: The learning rate satisfies $\eta_t \leq \frac{1}{L_H}$ for all t.
(A4) Hessian Approximation Error: The Hessian approximation error is bounded: $\|H^{[t]} - H_{-j}^{[t]}\| \leq \epsilon_H$, $\forall t$, where $H_{-j}^{[t]} = \frac{1}{|S_t\setminus\{j\}|}\sum_{i\in S_t\setminus\{j\}}\nabla^2\ell(z_i;\theta^{[t]})$ is the empirical Hessian over the mini-batch.
(A5) Gradient Norm Bound: For all $\theta \in \Theta$ and all z_i: $\|\nabla\ell(z_i;\theta)\| \leq G$.
(A6) Parameter Difference Bound: There exists a constant $M > 0$ such that: $\|\theta_{-j}^{[t]} - \theta^{[t]}\| \leq M$, $\forall t \in [t_1, t_2]$.
(A7) Bounded Hessian Norm: For all $\theta \in \Theta$ and all z_i: $\|\nabla^2\ell(z_i;\theta)\| \leq M_H$.

Then, the expected estimation error is bounded as follows:

$$\mathbb{E}\left[\left\|\Delta\theta_{-j}^{[t_1,t_2]} - \widehat{\Delta\theta}_{-j}^{[t_1,t_2]}\right\|\right] \leq \frac{\tilde{B}}{M_H}\left(e^{M_H\eta_{\max}(t_2+1)} + e^{M_H\eta_{\max}(t_1+1)} - 2\right) \tag{4.23}$$

where $\eta_{\max} = \max_{t\in[t_1,t_2]}\eta_t$ and $\tilde{B} = \frac{L_H M^2}{2} + \epsilon_H M$.

Proof **Step 1: Derivation of the Error Update Equation**
Define the error at iteration t:

$$e^{[t]} = (\theta_{-j}^{[t]} - \theta^{[t]}) - \widehat{\Delta\theta}_{-j}^{[t]} \tag{4.24}$$

where $\widehat{\Delta\theta}_{-j}^{[t]} = \widehat{\Delta\theta}_{-j}^{[0,t]}$ is the approximation of the true parameter change $\Delta\theta_{-j}^{[t]}$ using the TVIM method.

Our aim is to derive a recursive relation for $e^{[t]}$ and then bound its expected norm.
Consider the updates for $\theta^{[t]}$, $\theta_{-j}^{[t]}$, and $\widehat{\theta}_{-j}^{[t]}$:
Original SGD Update:

$$\theta^{[t+1]} = \theta^{[t]} - \eta_t\tilde{g}^{[t]}, \quad \tilde{g}^{[t]} = \frac{1}{|S_t|}\sum_{i\in S_t}\nabla\ell(z_i;\theta^{[t]}). \tag{4.25}$$

Leave-One-Out SGD Update:

$$\theta_{-j}^{[t+1]} = \theta_{-j}^{[t]} - \eta_t \tilde{g}_{-j}^{[t]}, \quad \tilde{g}_{-j}^{[t]} = \frac{1}{|S_t|} \sum_{i \in S_t \setminus \{j\}} \nabla \ell(z_i; \theta_{-j}^{[t]}). \tag{4.26}$$

Approximate Leave-One-Out Update (TVIM):

$$\widehat{\theta}_{-j}^{[t+1]} = \widehat{\theta}_{-j}^{[t]} - \eta_t \left(\tilde{g}^{[t]} + H^{[t]}(\widehat{\theta}_{-j}^{[t]} - \theta^{[t]}) - \mathbf{1}_{\{j \in S_t\}} \frac{1}{|S_t|} \nabla \ell(z_j; \theta^{[t]}) \right). \tag{4.27}$$

We derive the error update equation as follows:

$$e^{[t]} - e^{[t-1]} = \eta_{t-1} \delta^{[t-1]}, \tag{4.28}$$

where:

$$\delta^{[t-1]} = \left(\tilde{g}_{-j}^{[t-1]} - \tilde{g}^{[t-1]} \right) - H^{[t-1]} \widehat{\Delta\theta}_{-j}^{[t-1]} + \mathbf{1}_{\{j \in S_{t-1}\}} \frac{1}{|S_{t-1}|} \nabla \ell(z_j; \theta^{[t-1]}). \tag{4.29}$$

or equivalently:

$$\delta^{[t]} = \left(\tilde{g}_{-j}^{[t]} - \tilde{g}^{[t]} \right) - H^{[t]} \widehat{\Delta\theta}_{-j}^{[t]} + \mathbf{1}_{\{j \in S_t\}} \frac{1}{|S_t|} \nabla \ell(z_j; \theta^{[t]}). \tag{4.30}$$

Step 2: Bounding $\|\delta^{[t]}\|$

We decompose $\delta^{[t]}$ and bound each term.

1. Difference in Stochastic Gradients:

$$\tilde{g}_{-j}^{[t]} - \tilde{g}^{[t]} = \frac{1}{|S_t|} \left(\sum_{i \in S_t \setminus \{j\}} \left(\nabla \ell(z_i; \theta_{-j}^{[t]}) - \nabla \ell(z_i; \theta^{[t]}) \right) - \mathbf{1}_{\{j \in S_t\}} \nabla \ell(z_j; \theta^{[t]}) \right). \tag{4.31}$$

Applying a first-order Taylor expansion to $\nabla \ell(z_i; \theta_{-j}^{[t]})$ for $i \neq j$:

$$\nabla \ell(z_i; \theta_{-j}^{[t]}) - \nabla \ell(z_i; \theta^{[t]}) = \nabla^2 \ell(z_i; \theta^{[t]})(\theta_{-j}^{[t]} - \theta^{[t]}) + r_{i,j}^{[t]}, \tag{4.32}$$

where, by Assumption (A2):

$$\|r_{i,j}^{[t]}\| \leq \frac{L_H}{2} \|\theta_{-j}^{[t]} - \theta^{[t]}\|^2. \tag{4.33}$$

Thus, we have:

$$
\begin{aligned}
\tilde{g}_{-j}^{[t]} - \tilde{g}^{[t]} &= \frac{1}{|S_t|} \sum_{i \in S_t \setminus \{j\}} \nabla^2 \ell(z_i; \theta^{[t]})(\theta_{-j}^{[t]} - \theta^{[t]}) + r_{i,j}^{[t]} - \mathbf{1}_{\{j \in S_t\}} \nabla \ell(z_j; \theta^{[t]}) \\
&= \frac{1}{|S_t|} \left(\sum_{i \in S_t \setminus \{j\}} r_{i,j}^{[t]} - \mathbf{1}_{\{j \in S_t\}} \nabla \ell(z_j; \theta^{[t]}) \right) + H_{-j}^{[t]} (\theta_{-j}^{[t]} - \theta^{[t]}) \quad (4.34)
\end{aligned}
$$

2. Hessian Approximation Error:

$$\|(H_{-j}^{[t]} - H^{[t]})(\theta_{-j}^{[t]} - \theta^{[t]})\| \le \epsilon_H \|\theta_{-j}^{[t]} - \theta^{[t]}\|. \quad (4.35)$$

according to Assumption (A4).

3. Combining Terms: Substitute the approximations back into $\delta^{[t]}$:

$$
\begin{aligned}
\delta^{[t]} &= \left(\tilde{g}_{-j}^{[t]} - \tilde{g}^{[t]} \right) - H^{[t]} \widehat{\Delta\theta}_{-j}^{[t]} + \mathbf{1}_{\{j \in S_t\}} \frac{1}{|S_t|} \nabla \ell(z_j; \theta^{[t]}) \\
&= \left(\tilde{g}_{-j}^{[t]} - \tilde{g}^{[t]} \right) - H_{-j}^{[t]} (\theta_{-j}^{[t]} - \theta^{[t]}) \\
&\quad + \left(H_{-j}^{[t]} - H^{[t]} \right) (\theta_{-j}^{[t]} - \theta^{[t]}) + \mathbf{1}_{\{j \in S_t\}} \frac{1}{|S_t|} \nabla \ell(z_j; \theta^{[t]}) \\
&= \frac{1}{|S_t|} \sum_{i \in S_t \setminus \{j\}} r_{i,j}^{[t]} + (H_{-j}^{[t]} - H^{[t]})(\theta_{-j}^{[t]} - \theta^{[t]}) + H^{[t]} \big((\theta_{-j}^{[t]} - \theta^{[t]}) - \widehat{\Delta\theta}_{-j}^{[t]} \big) \\
&= \frac{1}{|S_t|} \sum_{i \in S_t \setminus \{j\}} r_{i,j}^{[t]} + (H_{-j}^{[t]} - H^{[t]})(\theta_{-j}^{[t]} - \theta^{[t]}) + H^{[t]} e^{[t]}. \quad (4.36)
\end{aligned}
$$

4. Bounding $\|\delta^{[t]}\|$: For the first term, $\left\| \frac{1}{|S_t|} \sum_{i \in S_t \setminus \{j\}} r_{i,j}^{[t]} \right\| < \frac{L_H M^2}{2}$. For the second term, $\left\| (H_{-j}^{[t]} - H^{[t]})(\theta_{-j}^{[t]} - \theta^{[t]}) \right\| \le \epsilon_H M$. For the third term, $\left\| H^{[t]} e^{[t]} \right\| \le M_H \|e^{[t]}\|$.

Combining bounds, we obtain:

$$\|\delta^{[t]}\| < \frac{L_H M^2}{2} + \epsilon_H M + M_H \|e^{[t]}\|. \quad (4.37)$$

Step 3: Error Update Equation

Using the error update:

$$e^{[t]} = e^{[t-1]} - \eta_t \delta^{[t-1]}, \quad (4.38)$$

we have:

$$\|e^{[t]}\| \leq \|e^{[t-1]}\| + \eta_t \|\delta^{[t-1]}\| < \|e^{[t-1]}\| + \eta_t \left(\frac{L_H M^2}{2} + \epsilon_H M + M_H \|e^{[t-1]}\| \right). \tag{4.39}$$

Define:

$$a_t = 1 + \eta_t M_H, \quad b_t = \eta_t \left(\frac{L_H M^2}{2} + \epsilon_H M \right). \tag{4.40}$$

Then:

$$\|e^{[t]}\| < a_t \|e^{[t-1]}\| + b_t. \tag{4.41}$$

Step 4: Taking Expectations

Taking expectations over the mini-batch sampling:

$$\mathbb{E}\left[\|e^{[t]}\|\right] < a_t \mathbb{E}\left[\|e^{[t-1]}\|\right] + b_t. \tag{4.42}$$

Define:

$$\tilde{B} = \frac{L_H M^2}{2} + \epsilon_H M. \tag{4.43}$$

Then:

$$\mathbb{E}\left[\|e^{[t]}\|\right] < a_t \mathbb{E}\left[\|e^{[t-1]}\|\right] + \eta_t \tilde{B}. \tag{4.44}$$

Step 5: Solving the Recurrence Relation

Unfolding the recurrence:

$$\mathbb{E}\left[\|e^{[t]}\|\right] \leq \prod_{k=0}^{t} a_k \cdot \mathbb{E}\left[\|e^{[0]}\|\right] + \sum_{s=0}^{t} \left(\prod_{k=s+1}^{t} a_k \right) b_s. \tag{4.45}$$

Since $e^{[0]} = 0$, we have:

$$\mathbb{E}\left[\|e^{[t]}\|\right] \leq \sum_{s=0}^{t} \left(\prod_{k=s+1}^{t} a_k \right) b_s. \tag{4.46}$$

Assuming $a_k \leq e^{M_H \eta_{\max}}$, we get:

$$\prod_{k=s+1}^{t} a_k \leq e^{M_H \eta_{\max}(t-s)}. \tag{4.47}$$

Therefore:

$$\mathbb{E}\left[\|e^{[t]}\|\right] \leq \tilde{B} \eta_{\max} \sum_{s=0}^{t} e^{M_H \eta_{\max}(t-s)}. \tag{4.48}$$

Approximating the sum:

$$\mathbb{E}\left[\|e^{[t]}\|\right] \leq \tilde{B}\eta_{\max} \cdot \frac{e^{M_H \eta_{\max}(t+1)} - 1}{e^{M_H \eta_{\max}} - 1}. \tag{4.49}$$

For small $M_H \eta_{\max}$, $e^{M_H \eta_{\max}} - 1 \approx M_H \eta_{\max}$, yielding:

$$\mathbb{E}\left[\|e^{[t]}\|\right] \leq \frac{\tilde{B}}{M_H}\left(e^{M_H \eta_{\max}(t+1)} - 1\right). \tag{4.50}$$

Substitute t with t_1 and t_2 respectively:

$$\mathbb{E}\left[\|e^{[t_2]}\|\right] \leq \frac{\tilde{B}}{M_H}\left(e^{M_H \eta_{\max}(t_2+1)} - 1\right), \tag{4.51}$$

$$\mathbb{E}\left[\|e^{[t_1]}\|\right] \leq \frac{\tilde{B}}{M_H}\left(e^{M_H \eta_{\max}(t_1+1)} - 1\right). \tag{4.52}$$

Step 6: Final Bound

The estimation error is:

$$\begin{aligned}\mathbb{E}\left[\left\|\Delta\theta_{-j}^{[t_1,t_2]} - \widehat{\Delta\theta}_{-j}^{[t_1,t_2]}\right\|\right] &\leq \mathbb{E}\left[\|e^{[t_2]}\|\right] + \mathbb{E}\left[\|e^{[t_1]}\|\right] \\ &\leq \frac{\tilde{B}}{M_H}\left(e^{M_H \eta_{\max}(t_2+1)} + e^{M_H \eta_{\max}(t_1+1)} - 2\right)\end{aligned} \tag{4.53}$$

This completes the proof. □

Remark 4.1 The error bound provides several key insights. The error grows at most exponentially with both t_1 and t_2, highlighting the challenge of long-range influence estimation; the impact of t_2 is generally more significant as it represents the end of the time window. The Hessian approximation error ϵ_H directly impacts the overall error, emphasizing the importance of accurate Hessian estimation. The maximum learning rate $\eta_{\max}$ affects the error bound exponentially, suggesting that smaller learning rates might help control the estimation error. Finally, the bound depends on the Lipschitz constants of the gradient and Hessian (L_g and L_H), indicating that smoother loss landscapes lead to more reliable influence estimates. This theorem provides a theoretical foundation for understanding the limitations of influence estimation without assuming convexity and guides practical considerations in its application to large-scale machine learning problems.

Time-Varying Influence Toolkit

The query-based formulation of TVIM can be instantiated with different query vectors to investigate specific aspects of model behavior. This subsection provides a toolkit of four such instantiations, covering loss value, prediction changes, feature importance, and parameter importance. Each is stated as a theorem with a complete proof.

TVIM for Loss Value

Theorem 4.2 (TVIM for Loss Value) *Given a loss function $\ell(z;\theta)$, a time window $[t_1, t_2]$, a training data point z_j, and a test data point z_{test}, the query-based influence measurement with query function $q(t) = \nabla_\theta \ell(z_{test}; \theta^{[t]})$ can be approximated as:*

$$Q_{-j}^{[t_1,t_2]}(q) \approx [\ell(z_{test}; \theta_{-j}^{[t_2]}) - \ell(z_{test}; \theta_{-j}^{[t_1]})] - [\ell(z_{test}; \theta^{[t_2]}) - \ell(z_{test}; \theta^{[t_1]})], \quad (4.54)$$

where $\theta_{-j}^{[t]}$ denotes the model parameters at time t when trained without data point z_j, and $\theta^{[t]}$ denotes the parameters when trained with all data points.

Proof We begin with the definition of the query-based influence measurement:

$$Q_{-j}^{[t_1,t_2]}(q) = \left\langle q(t_2), \Delta\theta_{-j}^{[t_2]} \right\rangle - \left\langle q(t_1), \Delta\theta_{-j}^{[t_1]} \right\rangle \quad (4.55)$$

where $\Delta\theta_{-j}^{[t]} = \theta_{-j}^{[t]} - \theta^{[t]}$.

Substituting $q(t) = \nabla_\theta \ell(z_{\text{test}}; \theta^{[t]})$ into Eq. (4.55):

$$Q_{-j}^{[t_1,t_2]}(q) = \left\langle \nabla_\theta \ell(z_{\text{test}}; \theta^{[t_2]}), \theta_{-j}^{[t_2]} - \theta^{[t_2]} \right\rangle - \left\langle \nabla_\theta \ell(z_{\text{test}}; \theta^{[t_1]}), \theta_{-j}^{[t_1]} - \theta^{[t_1]} \right\rangle. \quad (4.56)$$

Apply the first-order Taylor expansion of $\ell(z_{\text{test}}; \theta)$ around $\theta^{[t_2]}$ and $\theta^{[t_1]}$:

$$\ell(z_{\text{test}}; \theta_{-j}^{[t_2]}) \approx \ell(z_{\text{test}}; \theta^{[t_2]}) + \langle \nabla_\theta \ell(z_{\text{test}}; \theta^{[t_2]}), \theta_{-j}^{[t_2]} - \theta^{[t_2]} \rangle \quad (4.57)$$

$$\ell(z_{\text{test}}; \theta_{-j}^{[t_1]}) \approx \ell(z_{\text{test}}; \theta^{[t_1]}) + \langle \nabla_\theta \ell(z_{\text{test}}; \theta^{[t_1]}), \theta_{-j}^{[t_1]} - \theta^{[t_1]} \rangle \quad (4.58)$$

Rearranging Eqs. (4.57) and (4.58):

$$\langle \nabla_\theta \ell(z_{\text{test}}; \theta^{[t_2]}), \theta_{-j}^{[t_2]} - \theta^{[t_2]} \rangle \approx \ell(z_{\text{test}}; \theta_{-j}^{[t_2]}) - \ell(z_{\text{test}}; \theta^{[t_2]}) \quad (4.59)$$

$$\langle \nabla_\theta \ell(z_{\text{test}}; \theta^{[t_1]}), \theta_{-j}^{[t_1]} - \theta^{[t_1]} \rangle \approx \ell(z_{\text{test}}; \theta_{-j}^{[t_1]}) - \ell(z_{\text{test}}; \theta^{[t_1]}) \quad (4.60)$$

Substituting these approximations back into Eq. (4.56):

$$
\begin{aligned}
Q_{-j}^{[t_1,t_2]}(q) &\approx [\ell(z_{\text{test}}; \theta_{-j}^{[t_2]}) - \ell(z_{\text{test}}; \theta^{[t_2]})] - [\ell(z_{\text{test}}; \theta_{-j}^{[t_1]}) - \ell(z_{\text{test}}; \theta^{[t_1]})] \\
&= [\ell(z_{\text{test}}; \theta_{-j}^{[t_2]}) - \ell(z_{\text{test}}; \theta_{-j}^{[t_1]})] - [\ell(z_{\text{test}}; \theta^{[t_2]}) - \ell(z_{\text{test}}; \theta^{[t_1]})] \quad (4.61)
\end{aligned}
$$

This completes the proof of Theorem 4.2. □

This theorem provides a foundation for understanding how individual training data points affect the model's loss on specific test data points over time. The right-hand side of Eq. (4.54) represents the difference between the loss changes with and without data point z_j, offering a direct measure of that data point's influence on model performance.

Extension to Test Sets. This concept extends naturally to an entire test set $\mathcal{D}_{\text{test}} = \{z_1, \ldots, z_M\}$. Define the query function as:

$$
q(t) = \frac{1}{M} \sum_{i=1}^{M} \nabla_\theta \ell(z_i; \theta^{[t]}), \quad z_i \in \mathcal{D}_{\text{test}}. \tag{4.62}
$$

With this choice, the influence measurement approximates the change in average test loss:

$$
\begin{aligned}
Q_{-j}^{[t_1,t_2]}(q) &\approx \frac{1}{M} \sum_{i=1}^{M} \left[\ell(z_i; \theta_{-j}^{[t_2]}) - \ell(z_i; \theta_{-j}^{[t_1]})\right] - \frac{1}{M} \sum_{i=1}^{M} [\ell(z_i; \theta^{[t_2]}) - \ell(z_i; \theta^{[t_1]})] \\
&= \left[\mathcal{L}_{\text{test}}(\theta_{-j}^{[t_2]}) - \mathcal{L}_{\text{test}}(\theta_{-j}^{[t_1]})\right] - [\mathcal{L}_{\text{test}}(\theta^{[t_2]}) - \mathcal{L}_{\text{test}}(\theta^{[t_1]})],
\end{aligned} \tag{4.63}
$$

where $\mathcal{L}_{\text{test}}(\theta^{[t]}) = \frac{1}{M} \sum_{i=1}^{M} \ell(z_i; \theta^{[t]})$ is the average test loss.

TVIM for Prediction Changes

Theorem 4.3 (TVIM for Prediction Changes) *Given a model function $f(x; \theta)$, a time window $[t_1, t_2]$, a training data point z_j, and a test input x_{test}, the query-based influence measurement with query function $q(t) = \nabla_\theta f(x_{test}; \theta^{[t]})$ can be approximated as:*

$$
Q_{-j}^{[t_1,t_2]}(q) \approx \left[f(x_{test}; \theta_{-j}^{[t_2]}) - f(x_{test}; \theta_{-j}^{[t_1]})\right] - [f(x_{test}; \theta^{[t_2]}) - f(x_{test}; \theta^{[t_1]})], \tag{4.64}
$$

where $\theta_{-j}^{[t]}$ denotes the model parameters at time t when trained without data point z_j, and $\theta^{[t]}$ denotes the parameters when trained with all data points.

Proof We begin with the definition of the query-based influence measurement:

$$Q_{-j}^{[t_1,t_2]}(q) = \left\langle q(t_2), \Delta\theta_{-j}^{[t_2]} \right\rangle - \left\langle q(t_1), \Delta\theta_{-j}^{[t_1]} \right\rangle \tag{4.65}$$

where $\Delta\theta_{-j}^{[t]} = \theta_{-j}^{[t]} - \theta^{[t]}$.

Substituting $q(t) = \nabla_\theta f(x_{\text{test}}; \theta^{[t]})$ into Eq. (4.65):

$$Q_{-j}^{[t_1,t_2]}(q) = \left\langle \nabla_\theta f(x_{\text{test}}; \theta^{[t_2]}), \theta_{-j}^{[t_2]} - \theta^{[t_2]} \right\rangle - \left\langle \nabla_\theta f(x_{\text{test}}; \theta^{[t_1]}), \theta_{-j}^{[t_1]} - \theta^{[t_1]} \right\rangle. \tag{4.66}$$

We apply the first-order Taylor approximation of the model function around $\theta^{[t_2]}$ and $\theta^{[t_1]}$:

$$f(x_{\text{test}}; \theta_{-j}^{[t_2]}) \approx f(x_{\text{test}}; \theta^{[t_2]}) + \langle \nabla_\theta f(x_{\text{test}}; \theta^{[t_2]}), \theta_{-j}^{[t_2]} - \theta^{[t_2]} \rangle \tag{4.67}$$

$$f(x_{\text{test}}; \theta_{-j}^{[t_1]}) \approx f(x_{\text{test}}; \theta^{[t_1]}) + \langle \nabla_\theta f(x_{\text{test}}; \theta^{[t_1]}), \theta_{-j}^{[t_1]} - \theta^{[t_1]} \rangle \tag{4.68}$$

Rearranging these equations:

$$\langle \nabla_\theta f(x_{\text{test}}; \theta^{[t_2]}), \theta_{-j}^{[t_2]} - \theta^{[t_2]} \rangle \approx f(x_{\text{test}}; \theta_{-j}^{[t_2]}) - f(x_{\text{test}}; \theta^{[t_2]}) \tag{4.69}$$

$$\langle \nabla_\theta f(x_{\text{test}}; \theta^{[t_1]}), \theta_{-j}^{[t_1]} - \theta^{[t_1]} \rangle \approx f(x_{\text{test}}; \theta_{-j}^{[t_1]}) - f(x_{\text{test}}; \theta^{[t_1]}) \tag{4.70}$$

Substituting these approximations back into Eq. (4.66):

$$\begin{aligned} Q_{-j}^{[t_1,t_2]}(q) &\approx [f(x_{\text{test}}; \theta_{-j}^{[t_2]}) - f(x_{\text{test}}; \theta^{[t_2]})] - [f(x_{\text{test}}; \theta_{-j}^{[t_1]}) - f(x_{\text{test}}; \theta^{[t_1]})] \\ &= [f(x_{\text{test}}; \theta_{-j}^{[t_2]}) - f(x_{\text{test}}; \theta_{-j}^{[t_1]})] - [f(x_{\text{test}}; \theta^{[t_2]}) - f(x_{\text{test}}; \theta^{[t_1]})] \end{aligned} \tag{4.71}$$

This completes the proof of Theorem 4.3. □

This theorem provides a formal justification for using TVIM to analyze how excluding data point z_j influences the model's predictions on a test input x_{test} over the interval $[t_1, t_2]$. Compared to Theorem 4.2, which focuses on the loss value, Theorem 4.3 focuses on specific model outputs. It enables the identification of influential training data points for specific predictions, aids in understanding model behavior on particular inputs, and can help detect potential outliers or mislabeled data points.

TVIM for Feature Importance

Theorem 4.4 (TVIM for Feature Importance) *Given a loss function* $\ell(z; \theta)$, *a training data point* $z = (x, y)$, *and a test data point* $z_{test} = (x_{test}, y_{test})$, *the query-based influence measurement for feature importance with query function* $q(t) = \nabla_x \nabla_\theta \ell(z_{test}; \theta^{[t]})$ *can be approximated as:*

$$Q_{-j}^{[t_1,t_2]}(q) \approx [\nabla_x \ell(z_{test}; \theta_{-j}^{[t_2]}) - \nabla_x \ell(z_{test}; \theta_{-j}^{[t_1]})] - [\nabla_x \ell(z_{test}; \theta^{[t_2]}) - \nabla_x \ell(z_{test}; \theta^{[t_1]})], \tag{4.72}$$

where $\theta_{-j}^{[t]}$ denotes the model parameters at time t when trained without data point z_j, and $\theta^{[t]}$ denotes the parameters when trained with all data points.

Proof We start with the definition of the query-based influence measurement:

$$Q_{-j}^{[t_1,t_2]}(q) = \left\langle q(t_2), \Delta\theta_{-j}^{[t_2]} \right\rangle - \left\langle q(t_1), \Delta\theta_{-j}^{[t_1]} \right\rangle, \tag{4.73}$$

where $\Delta\theta_{-j}^{[t]} = \theta_{-j}^{[t]} - \theta^{[t]}$.

Substituting $q(t) = \nabla_x \nabla_\theta \ell(z_{\text{test}}; \theta^{[t]})$:

$$Q_{-j}^{[t_1,t_2]}(q) = \left\langle \nabla_\theta \nabla_x \ell(z_{\text{test}}; \theta^{[t_2]}), \theta_{-j}^{[t_2]} - \theta^{[t_2]} \right\rangle - \left\langle \nabla_\theta \nabla_x \ell(z_{\text{test}}; \theta^{[t_1]}), \theta_{-j}^{[t_1]} - \theta^{[t_1]} \right\rangle. \tag{4.74}$$

We apply the first-order Taylor approximation of $\nabla_x \ell(z_{\text{test}}; \theta)$ around $\theta^{[t_2]}$ and $\theta^{[t_1]}$:

$$\nabla_x \ell(z_{\text{test}}; \theta_{-j}^{[t_2]}) \approx \nabla_x \ell(z_{\text{test}}; \theta^{[t_2]}) + \nabla_\theta \nabla_x \ell(z_{\text{test}}; \theta^{[t_2]}) \left(\theta_{-j}^{[t_2]} - \theta^{[t_2]}\right), \tag{4.75}$$

$$\nabla_x \ell(z_{\text{test}}; \theta_{-j}^{[t_1]}) \approx \nabla_x \ell(z_{\text{test}}; \theta^{[t_1]}) + \nabla_\theta \nabla_x \ell(z_{\text{test}}; \theta^{[t_1]}) \left(\theta_{-j}^{[t_1]} - \theta^{[t_1]}\right). \tag{4.76}$$

Rearranging these equations:

$$\left\langle \nabla_\theta \nabla_x \ell(z_{\text{test}}; \theta^{[t_2]}), \theta_{-j}^{[t_2]} - \theta^{[t_2]} \right\rangle \approx \nabla_x \ell(z_{\text{test}}; \theta_{-j}^{[t_2]}) - \nabla_x \ell(z_{\text{test}}; \theta^{[t_2]}), \tag{4.77}$$

$$\left\langle \nabla_\theta \nabla_x \ell(z_{\text{test}}; \theta^{[t_1]}), \theta_{-j}^{[t_1]} - \theta^{[t_1]} \right\rangle \approx \nabla_x \ell(z_{\text{test}}; \theta_{-j}^{[t_1]}) - \nabla_x \ell(z_{\text{test}}; \theta^{[t_1]}). \tag{4.78}$$

Substituting these approximations back into Eq. (4.74):

$$\begin{aligned} Q_{-j}^{[t_1,t_2]}(q) &\approx \left[\nabla_x \ell(z_{\text{test}}; \theta_{-j}^{[t_2]}) - \nabla_x \ell(z_{\text{test}}; \theta^{[t_2]})\right] - \left[\nabla_x \ell(z_{\text{test}}; \theta_{-j}^{[t_1]}) - \nabla_x \ell(z_{\text{test}}; \theta^{[t_1]})\right] \\ &= \left[\nabla_x \ell(z_{\text{test}}; \theta_{-j}^{[t_2]}) - \nabla_x \ell(z_{\text{test}}; \theta_{-j}^{[t_1]})\right] - \left[\nabla_x \ell(z_{\text{test}}; \theta^{[t_2]}) - \nabla_x \ell(z_{\text{test}}; \theta^{[t_1]})\right]. \end{aligned} \tag{4.79}$$

This completes the proof. □

This theorem shows how TVIM measures the impact of excluding a training data point z_j on the gradient of the loss with respect to the input features at the test data point z_{test} over the interval $[t_1, t_2]$. This provides insights into how the importance of different input features evolves during training and how individual training data points influence this feature importance.

TVIM for Parameter Importance

Theorem 4.5 (TVIM for Parameter Importance) *Given a model with parameters $\theta \in \mathbb{R}^p$, a time window $[t_1, t_2]$, a training data point z_j, and the i-th standard basis vector $e_i \in \mathbb{R}^p$, the query-based influence measurement with query function $q(t) = e_i$ is exactly:*

$$Q_{-j}^{[t_1,t_2]}(q) = \left(\theta_{-j,i}^{[t_2]} - \theta_{-j,i}^{[t_1]}\right) - \left(\theta_i^{[t_2]} - \theta_i^{[t_1]}\right), \tag{4.80}$$

where $\theta_{-j,i}^{[t]}$ denotes the i-th component of the model parameters at time t when trained without data point z_j, and $\theta_i^{[t]}$ denotes the i-th component of the parameters when trained with all data points.

Proof We start with the definition of the query-based influence measurement:

$$Q_{-j}^{[t_1,t_2]}(q) = \left\langle q(t_2), \Delta\theta_{-j}^{[t_2]}\right\rangle - \left\langle q(t_1), \Delta\theta_{-j}^{[t_1]}\right\rangle, \tag{4.81}$$

where $\Delta\theta_{-j}^{[t]} = \theta_{-j}^{[t]} - \theta^{[t]}$.

Substituting $q(t) = e_i$, which is constant over time:

$$Q_{-j}^{[t_1,t_2]}(q) = \left\langle e_i, \theta_{-j}^{[t_2]} - \theta^{[t_2]}\right\rangle - \left\langle e_i, \theta_{-j}^{[t_1]} - \theta^{[t_1]}\right\rangle. \tag{4.82}$$

Since e_i is the i-th standard basis vector, the inner product selects the i-th component:

$$Q_{-j}^{[t_1,t_2]}(q) = \left(\theta_{-j,i}^{[t_2]} - \theta_i^{[t_2]}\right) - \left(\theta_{-j,i}^{[t_1]} - \theta_i^{[t_1]}\right) = \left(\theta_{-j,i}^{[t_2]} - \theta_{-j,i}^{[t_1]}\right) - \left(\theta_i^{[t_2]} - \theta_i^{[t_1]}\right). \tag{4.83}$$

This matches the expression in Eq. (4.80), completing the proof. □

This theorem allows us to isolate the influence of a training data point z_j on specific model parameters over the interval $[t_1, t_2]$. A large absolute value of $Q_{-j}^{[t_1,t_2]}(q)$ indicates that data point z_j has a significant influence on the i-th parameter during the specified time window. This is particularly useful for identifying which parameters are most affected by specific training data points and understanding the localized effects of individual data points on the model.

By analyzing how $Q_{-j}^{[t_1,t_2]}(q)$ changes over different time windows, we can understand how the influence of a training data point on specific parameters evolves throughout the training process.

Experimental Evaluation

TVIM's theoretical guarantees hold under standard non-convex assumptions, but their practical relevance depends on whether the framework produces accurate and actionable influence estimates on real data and models. The experiments reported here address three questions: how accurately does TVIM estimate influence compared to existing methods; what can be learned by analyzing how influence evolves across various training stages; and whether time-window analysis of influence can improve practical machine learning tasks. Section "Experimental Setup" describes the experimental setup. Section "Accuracy of Influence Measurement" validates the accuracy of TVIM against influence functions using leave-one-out as ground truth. Section "Influence Dynamics and Similarity Across Training Stages" analyzes the similarity of influence rankings across training stages. Section "Applications of Time-Varying Influence Measurement" applies TVIM to the practical task of detecting flipped labels.

Experimental Setup

We evaluate TVIM across diverse datasets and model architectures, comparing it against leading influence estimation methods. Our experimental setup included eight Nvidia RTX A5000 GPUs, each equipped with 24 GB of memory. These were supported by dual Intel Xeon Gold 6342 CPUs running at 2.80 GHz with 96 cores in total and 503 GB of RAM. The software environment comprised Ubuntu 22.04.3 LTS (64-bit), PyTorch v2.4.1, CUDA 12.4, and Python 3.11.9. Code and data are available at https://github.com/dynamic-infl-tracker/DIT.

Datasets

We used four datasets spanning different domains and complexities: Adult (tabular) [16], 20Newsgroups (text) [17], MNIST [18] and EMNIST [19] (grayscale images).

Models

We used three model architectures of varying complexity. Logistic Regression (LR) is a simple linear model serving as a convex baseline. Deep Neural Network (DNN) has two hidden layers using ReLU activations. Convolutional Neural Network (CNN) has two convolutional layers followed by a fully connected layer. The DNN and CNN represent non-convex scenarios. All models are optimized using the binary cross-entropy loss with logits, which combines sigmoid activation with binary cross-entropy loss for binary classification tasks. Input and output dimensions were adapted to each dataset.

Comparison Methods

We evaluate TVIM against two established methods. Leave-One-Out (LOO) directly measures the influence of removing a training data point z_j by retraining the model:

$$\Delta\ell_{\text{LOO}}(z_j) = \frac{1}{M}\sum_{i=1}^{M}\big(\ell(z_i, \theta_{-j}) - \ell(z_i, \theta)\big), \tag{4.84}$$

where $z_i \in \mathcal{D}_{\text{test}}$ and M is the size of the test set $\mathcal{D}_{\text{test}} = \{z_i\}_{i=1}^{M}$. While LOO provides a robust ground truth baseline, it is computationally intensive. Influence Functions (IF) [1] estimate the influence of removing a training data point z_j on the model's overall loss for a test set $\mathcal{D}_{\text{test}}$:

$$I(z_j, \mathcal{D}_{\text{test}}) = -\frac{1}{M}\sum_{i=1}^{M}\nabla_\theta \ell(z_i, \theta)^T H^{-1} \nabla_\theta \ell(z_j, \theta), \tag{4.85}$$

where H is the Hessian of the model's loss at θ. For TVIM, we set the query function as $q(t) = \frac{1}{M}\sum_{i=1}^{M} \nabla_\theta \ell(z_i; \theta^{[t]})$, measuring the impact on the test set loss across the time window $[t_1, t_2]$:

$$Q_{-j}^{[t_1,t_2]}(q) \approx \frac{1}{M}\sum_{i=1}^{M}\big[\ell(z_i; \theta_{-j}^{[t_2]}) - \ell(z_i; \theta_{-j}^{[t_1]})\big] - \frac{1}{M}\sum_{i=1}^{M}\big[\ell(z_i; \theta^{[t_2]}) - \ell(z_i; \theta^{[t_1]})\big]. \tag{4.86}$$

To ensure the reproducibility and robustness of our results, we present them as the mean $\pm$ standard deviation calculated over 16 runs, each initialized with a different random seed.

Accuracy of Influence Measurement

Accurate influence estimation is the foundation of all downstream applications, from data valuation to unlearning. To validate TVIM's accuracy, we compared TVIM against IF using LOO as ground truth. We employed TVIM's full-time window $[0, T]$ for a fair comparison with IF, which can only measure overall data-point influence on the final model. To evaluate how closely TVIM and IF approximate LOO, we adopted four metrics: Pearson and Spearman correlations for linear and monotonic relationships, respectively, Kendall's tau for ordinal relationships, and Jaccard similarity for the top 30% influencers.

Table 4.1 reveals several key findings. TVIM consistently surpasses IF in accuracy across all datasets, model architectures, and evaluation metrics. TVIM's advantage is most significant in complex settings such as the non-convex DNN and the complex MNIST dataset. TVIM also shows superior robustness and reliability, with lower standard deviations across runs compared to IF.

Table 4.1 Performance comparison of TVIM and IF for logistic regression and deep neural network

Model	Dataset	Pearson		Spearman		Kendall's Tau		Jaccard	
		TVIM	IF	TVIM	IF	TVIM	IF	TVIM	IF
LR	Adult	**0.99 ± 0.01**	0.91 ± 0.04	**0.99 ± 0.01**	0.93 ± 0.02	**0.95 ± 0.01**	0.79 ± 0.04	**0.91 ± 0.04**	0.71 ± 0.06
	20News	**0.99 ± 0.01**	0.90 ± 0.13	**0.99 ± 0.01**	0.94 ± 0.08	**0.97 ± 0.01**	0.84 ± 0.13	**0.95 ± 0.03**	0.78 ± 0.16
	MNIST	**0.93 ± 0.10**	0.76 ± 0.14	**0.98 ± 0.01**	0.61 ± 0.22	**0.95 ± 0.02**	0.49 ± 0.21	**0.91 ± 0.05**	0.48 ± 0.14
DNN	Adult	**0.95 ± 0.02**	0.88 ± 0.04	**0.95 ± 0.03**	0.86 ± 0.04	**0.83 ± 0.06**	0.69 ± 0.05	**0.75 ± 0.08**	0.56 ± 0.07
	20News	**0.85 ± 0.07**	0.77 ± 0.05	**0.85 ± 0.08**	0.80 ± 0.06	**0.71 ± 0.08**	0.62 ± 0.07	**0.67 ± 0.08**	0.55 ± 0.07
	MNIST	**0.90 ± 0.07**	0.25 ± 0.28	**0.98 ± 0.01**	0.26 ± 0.33	**0.90 ± 0.03**	0.19 ± 0.24	**0.85 ± 0.05**	0.27 ± 0.19

Influence Dynamics and Similarity Across Training Stages

The accuracy results in the preceding subsection validate TVIM on the full training window, but the framework's distinctive capability is its ability to decompose influence into arbitrary sub-intervals. To exploit this capability, we analyzed the similarity of influence rankings across different training stages. The training process was adaptively divided into early, middle, and late stages using change points identified in the overall training loss trajectory. We then set time windows based on these stages and used TVIM to compute data-point influence within each window. We used Kendall's tau correlation to quantify the similarity of influence rankings between stages, with higher values indicating greater stability. Table 4.2 presents these correlations.

Table 4.2 reveals several key insights. Data-point influence evolves significantly throughout training, as evidenced by the consistently low correlations between early and late stages (Early-Late column); this challenges static influence measurement methods and highlights the necessity for time-aware methods like TVIM. Mid-training influence strongly correlates with full-training influence across all datasets and models, suggesting that influential data points can be identified before convergence; mid-training analysis may suffice for estimating full-training influence, potentially reducing computational costs, with significant implications for data selection and curriculum learning strategies. For a given dataset, the patterns of influence ranking changes at different stages are similar across different model architectures when accounting for standard deviations, suggesting that the influence of data points is largely determined by the inherent properties of the dataset rather than being heavily model-dependent.

Table 4.2 Kendall's Tau correlations across training stages

Model	Dataset	Early-middle	Early-late	Middle-late	Early-full	Middle-full	Late-full
LR	Adult	0.64 ± 0.14	0.62 ± 0.08	0.79 ± 0.14	0.81 ± 0.05	**0.82 ± 0.12**	0.79 ± 0.05
	20News	0.79 ± 0.12	0.78 ± 0.10	0.79 ± 0.09	**0.91 ± 0.02**	0.88 ± 0.10	0.86 ± 0.12
	MNIST	0.43 ± 0.14	0.15 ± 0.12	0.35 ± 0.14	0.71 ± 0.08	**0.72 ± 0.09**	0.30 ± 0.14
	EMNIST	0.73 ± 0.04	0.40 ± 0.16	0.51 ± 0.18	0.83 ± 0.03	**0.89 ± 0.02**	0.49 ± 0.17
DNN	Adult	0.61 ± 0.11	0.41 ± 0.15	0.70 ± 0.06	0.7 ± 0.09	**0.87 ±0.04**	0.69 ± 0.08
	20News	0.66 ± 0.06	0.57 ± 0.07	0.76 ± 0.05	0.81 ± 0.03	**0.82 ± 0.04**	0.76 ± 0.04
	MNIST	0.56 ± 0.06	0.18 ± 0.21	0.20 ± 0.25	0.74 ± 0.03	**0.81 ± 0.04**	0.20 ± 0.25
	EMNIST	0.60 ± 0.12	0.40 ± 0.20	0.59 ± 0.21	0.69 ± 0.11	**0.84 ± 0.07**	0.63 ± 0.17

Applications of Time-Varying Influence Measurement

Beyond characterizing influence dynamics, TVIM can be applied to practical data quality tasks that benefit from temporally resolved influence estimates. This subsection demonstrates one such application: detecting training data points with flipped labels.

Flipped-Label Detection

To demonstrate the practical utility of TVIM, we applied it to detecting flipped labels in a binary classification problem using the MNIST dataset (distinguishing between digits '1' and '7'). We randomly selected and flipped labels for 5%, 10%, 15%, and 20% of the training data, corresponding to 12, 25, 38, and 51 data points, respectively. Models were then trained on these partially corrupted datasets. We calculated influence using six methods: full-process TVIM, IF, LOO, and epoch-specific TVIM (first, middle, and last epochs). For each method, we ranked training data points by their negative influence and evaluated the top-k data points, where k equals the number of deliberately flipped data points. This approach allows us to assess each method's ability to identify mislabeled data points accurately. Table 4.3 presents results averaged over 16 runs.

The results show three clear findings. TVIM consistently outperforms IF across all scenarios, often matching or closely approaching the LOO performance, and maintains its performance advantage across varying levels of label noise (5% to 20%). The performance gap between TVIM and IF widens as model complexity increases

Table 4.3 Number of correctly identified flipped data points

Flipped	Model	IF	Full TVIM	LOO	First epoch	Mid epoch	Last epoch
5%	LR	10.50 ± 0.50	**10.94 ± 0.90**	**10.94 ± 0.90**	10.56 ± 1.22	10.88 ± 0.78	10.88 ± 0.78
	DNN	2.94 ± 2.01	9.06 ± 1.85	8.81 ± 1.98	8.25 ± 2.33	8.88 ± 2.09	**9.38 ± 1.98**
	CNN	5.88 ± 2.26	10.50 ± 1.32	10.44 ± 1.32	8.75 ± 2.11	10.69 ± 1.16	**11.06 ± 1.32**
10%	LR	23.44 ± 0.93	**23.50 ± 1.00**	**23.50 ± 1.00**	22.56 ± 1.54	**23.50 ± 1.06**	23.38 ± 1.00
	DNN	7.50 ± 3.34	20.75 ± 3.01	19.94 ± 3.77	20.31 ± 2.78	20.50 ± 3.22	**21.31 ± 3.77**
	CNN	15.00 ± 2.83	21.81 ± 3.11	21.75 ± 3.11	18.44 ± 4.37	22.19 ± 2.81	**23.56 ± 3.11**
15%	LR	**36.06 ± 0.97**	**36.06 ± 1.14**	**36.06 ± 1.14**	35.38 ± 1.62	35.69 ± 1.69	35.13 ± 1.14
	DNN	12.63 ± 4.62	32.81 ± 3.47	32.50 ± 3.72	32.19 ± 3.40	32.56 ± 3.61	**33.31 ± 3.72**
	CNN	23.44 ± 4.68	34.19 ± 4.17	34.19 ± 4.17	29.75 ± 5.93	34.56 ± 3.98	**36.31 ± 4.17**
20%	LR	48.63 ± 1.11	**48.69 ± 1.16**	**48.69 ± 1.16**	47.94 ± 1.52	46.56 ± 3.12	42.94 ± 1.16
	DNN	22.31 ± 6.14	45.31 ± 3.29	43.94 ± 5.20	44.13 ± 3.64	45.19 ± 3.30	**45.56 ± 5.20**
	CNN	31.00 ± 5.79	46.19 ± 4.33	46.25 ± 4.35	41.50 ± 7.66	47.13 ± 3.35	**48.69 ± 4.35**

(from LR to DNN to CNN), highlighting TVIM's robustness to non-convexity. Later training stages generally yield better detection accuracy, particularly for complex models; as models converge, the influence of mislabeled data points becomes more distinguishable relative to correctly labeled ones. These findings collectively demonstrate TVIM's effectiveness as a powerful tool for enhancing model robustness and data quality assessment, particularly in complex, real-world machine learning scenarios.

The flipped-label detection experiment confirms that TVIM provides practically useful influence estimates that translate into superior data quality diagnostics compared to IF, especially in non-convex settings. Notably, the finding that late-epoch TVIM often outperforms full-trajectory analysis for label detection suggests that targeted window selection can improve both efficiency and effectiveness.

Conclusion

This chapter extended the theory of data influence to a time-varying setting, providing a cumulative and temporally localized understanding of how data points shape model parameters across different training stages. This perspective captures the history of each data point's effect, showing not only whether but also when and to what extent its influence accumulates.

From the standpoint of machine unlearning theory, this formulation bridges the gap between instantaneous influence detection and practical forgetting. By decomposing the learning trajectory into time windows, one can identify the specific intervals during which a data point exerted the strongest influence and selectively target those intervals for correction or removal. The two-phase estimation framework, recording trajectory information once and replaying it later, makes this analysis computationally feasible, while theoretical error bounds ensure reliable estimation even in non-convex learning scenarios.

Together, the online and time-varying perspectives provide a unified theoretical basis for unlearning. The first enables real-time detection and prevention of harmful data influence, and the second supports correction of accumulated effects. These complementary frameworks together advance the theoretical foundations of machine unlearning by connecting data influence measurement with both proactive and reactive forgetting mechanisms.

References

1. Koh, P.W., Liang, P.: Understanding black-box predictions via influence functions. In: International Conference on Machine Learning. PMLR, pp. 1885–1894 (2017)
2. Ghorbani, A., Zou, J.: Data Shapley: equitable valuation of data for machine learning. In: International Conference on Machine Learning. PMLR, pp. 2242–2251 (2019)
3. Pruthi, G., et al.: Estimating training data influence by tracing gradient descent. Adv. Neural. Inf. Process. Syst. **33**, 19920–19930 (2020)
4. Shapley, L.S.: A value for n-person games. In: Contribution to the Theory of Games, vol. 2 (1953)
5. Toneva, M., et al.: An empirical study of example forgetting during deep neural network learning. In: International Conference on Learning Representations. https://openreview.net/forum?id=BJlxm30cKm (2019)
6. Swayamdipta, S., et al.: Dataset cartography: mapping and diagnosing datasets with training dynamics. In: Proceedings of the 2020 Conference on Empirical Methods in Natural Language Processing (EMNLP), pp. 9275–9293 (2020)
7. Paul, M., Ganguli, S., Dziugaite, G.K.: Deep learning on a data diet: finding important examples early in training. In: Advances in Neural Information Processing Systems, vol. 34, pp. 20596–20607 (2021)
8. Guo, C., et al.: Certified data removal from machine learning models. In: International Conference on Machine Learning. PMLR, pp. 3832–3842 (2020)
9. Neel, S., Roth, A., Sharifi-Malvajerdi, S.: Descent-to-delete: gradient-based methods for machine unlearning. In: Algorithmic Learning Theory. PMLR, pp. 931–962 (2021)
10. Thudi, A., et al.: Unrolling SGD: understanding factors influencing machine unlearning. In: IEEE 7th European Symposium on Security and Privacy (EuroS&P), pp. 303–319. IEEE (2022)
11. Hara, S., Nitanda, A., Maehara, T.: Data cleansing for models trained with SGD. In: Advances in Neural Information Processing Systems, vol. 32 (2019)
12. Nesterov, Y.: Introductory Lectures on Convex Optimization: A Basic Course. Springer (2004). https://doi.org/10.1007/978-1-4419-8853-9
13. Shalev-Shwartz, S., Ben-David, S.: Understanding Machine Learning: From Theory to Algorithms. Cambridge University Press (2014)
14. Robbins, H., Monro, S.: A stochastic approximation method. In: The Annals of Mathematical Statistics, pp. 400–407 (1951)
15. Bottou, L., Curtis, F.E., Nocedal, J.: Optimization methods for large-scale machine learning. SIAM Rev. **60**(2), 223–311 (2018)
16. Dua, D., Graff, C.: UCI Machine Learning Repository. http://archive.ics.uci.edu/ml (2019)

17. Lang, K.: Newsweeder: learning to filter netnews. In: International Conference on Machine Learning, pp. 331–339 (1995)
18. LeCun, Y., Cortes, C., Burges, C., et al.: MNIST handwritten digit database. http://yann.lecun.com/exdb/mnist (2010)
19. Cohen, G., et al.: EMNIST: extending MNIST to handwritten letters. In: Proceedings of the International Joint Conference on Neural Networks (2017)

Part II
Applications of Machine Unlearning

This part translates the theoretical tools of Part I into unlearning algorithms for largescale models and examines their application to wireless networks. It presents an exact unlearning method based on prompt tuning (Chap. 5), a certified approximate unlearning framework based on decay-aware noise injection (Chap. 6), and concludes with a discussion of open challenges and future directions for machine unlearning in networking systems (Chap. 7).

Chapter 5
Exact Unlearning for Large Models via Prompt Tuning

Abstract To address the growing demand for privacy protection in machine learning, we propose an efficient and exact machine unlearning method for **L**arge **M**odels, called **LM**Eraser. LMEraser takes a divide-and-conquer strategy with an adaptive prompt tuning mechanism to isolate data influence effectively. The training dataset is partitioned into public and private datasets. Public data are used to train the backbone of the model. Private data are clustered based on their diversity, and each cluster tunes a tailored prompt independently. This approach enables targeted unlearning by updating affected prompts, significantly reduces unlearning costs, and maintains high model performance. Evaluations show that LMEraser reduces unlearning costs by 100 times compared to prior work without compromising model utility.

Motivation and Contribution

Large models such as GPT [1], Vision Transformers (ViT) [2], and CLIP [3], have achieved remarkable performance across various tasks. However, this success has raised privacy concerns. Large models trained on massive datasets potentially memorize and expose sensitive user information [4]. To address these concerns, regulations [5, 6] grant users the "right to be forgotten." Consequently, machine unlearning, the ability to selectively remove the influence of specific training data from a trained model, has emerged as a key solution in balancing privacy protection and computational efficiency.

Existing unlearning methods, primarily designed for smaller-scale models, show inefficiency and ineffectiveness in addressing challenges posed by the scale and complexity of large models. For example, sharding-based unlearning methods [7, 8] partition training data randomly, training independent models on each shard and combining their outputs for predictions (Fig. 5.1). While unlearning only requires retraining the affected shard model, this method incurs high computational and storage costs when applied to large models, as it necessitates maintaining multiple large-scale shard models. Other methods attempt to manipulate parameters directly, but they rely on assumptions, including loss convexity, that do not hold for large models, failing to completely remove data influence [9–11].

J. Xu and X. Jia, *Machine Unlearning*, Wireless Networks,
https://doi.org/10.1007/978-3-032-30978-5_5

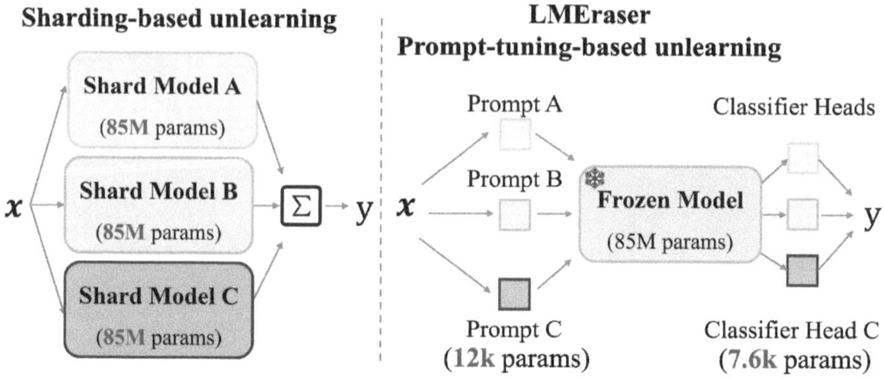

Fig. 5.1 Comparison of sharding-based unlearning and LMEraser. Sharding-based methods retrain entire shard models (about 85M parameters), while LMEraser efficiently updates affected prompts (about 12k parameters), drastically reducing computational costs

To completely remove the influence of specific training data from large models, there are several challenges. First, it is difficult to trace and quantify the impact of individual training data points, as their influence is dispersed across the model's architecture. Second, it is computationally expensive to implement unlearning operations, which involve intensive recalculations and adjustments to billions of parameters. Third, unlearning risks performance degradation and even catastrophic forgetting [12], as it may disrupt learned representations that are distributed across shared model components.

We propose LMEraser, an efficient and exact unlearning method that completely removes specific data influences from large vision models. The idea of LMEraser lies in two observations: (1) The training data of large models can be divided into two categories: *public data* and *private data*. Public data, such as ImageNet [13] and COCO [14], are collected, cleansed, and validated by research institutions or non-profits, making them widely available and extensively used. Private data, such as medical records or bank transactions, are sensitive and require permissions for access and use. While users (private data owners) may request removing their data from trained models due to the "right to be forgotten", such concerns are non-existent with public data. (2) Large models typically adopt a two-phase training process. The pre-training phase trains a model backbone using massive data to absorb broad foundational knowledge. Then, the tuning phase utilizes the backbone and task-specific data to enable the model to handle specific tasks.

Based on these foundations, LMEraser has three core designs to achieve large model unlearning. First, LMEraser takes a divide-and-conquer strategy with a prompt tuning architecture to isolate data influence, as shown in Fig. 5.1. The training dataset is partitioned into public and private datasets. Public data are used to pre-train the model's backbone, while private data are used for prompt tuning. Second, LMEraser freezes the backbone's parameters after pre-training, reducing recalculations during unlearning and preserving performance. Finally, LMEraser designs an adaptive mechanism for prompt tuning. It adaptively clusters private data based on data diversity and generates tailored prompts and classifier heads for each cluster. Predictions are made using the nearest cluster's prompt. Unlearning only requires updating the affected cluster's prompt and classifier head, reducing computational costs and minimizing performance degradation.

In summary, the contributions of this chapter include:

- We propose LMEraser, a novel unlearning method for large vision models that achieves complete and efficient removal of data influences while preserving model performance.
- We design an adaptive mechanism for prompt tuning that enhances performance and reduces unlearning costs. Our theoretical analysis highlights LMEraser's efficiency gains.
- LMEraser outperforms baselines and achieves a 100-fold reduction in unlearning costs while achieving high model utility.

System Model and Problem Statement

LMEraser operates on large-scale vision models and leverages visual prompt tuning to isolate private data influence into small, independently retrainable modules. This section introduces the two technical components that underpin LMEraser, namely the Vision Transformer architecture and the visual prompt tuning mechanism, and then formally defines exact unlearning and states the problem that LMEraser addresses.

System Model

The system model comprises two components: a Vision Transformer backbone that processes image inputs, and a visual prompt tuning mechanism that adapts the frozen backbone to downstream tasks without modifying its parameters.

Vision Transformer

The Vision Transformer (ViT) [15] splits an input image $\mathbf{x} \in \mathbb{R}^{H \times W \times C}$ into non-overlapping patches, where H, W, and C denote height, width, and number of channels, respectively. These patches are flattened into embeddings and combined with a learnable classification token to form a sequence $\mathbf{Z}$. The sequence $\mathbf{Z}$ is processed by a Transformer encoder [16], and the resulting output representation is fed to a classifier head that produces the final classification prediction.

Visual Prompt Tuning

Visual prompt tuning adapts a frozen pre-trained backbone to downstream tasks by learning a small input perturbation rather than updating the backbone parameters. The perturbation takes the form of a pixel-level border prompt, defined as follows.

Definition 5.1 (*Pixel-Level Border Prompt*) A pixel-level border prompt $\mathbf{p} \in \mathbb{R}^{H' \times W' \times C'}$ is a learnable modification overlaid on the borders of an image $\mathbf{x}$. The prompted image is obtained as

$$\tilde{\mathbf{x}} = g(\mathbf{x}, \mathbf{p}), \tag{5.1}$$

where $g \colon \mathbb{R}^{H \times W \times C} \times \mathbb{R}^{H' \times W' \times C'} \to \mathbb{R}^{H \times W \times C}$ overlays $\mathbf{p}$ onto $\mathbf{x}$. The prompted image $\tilde{\mathbf{x}}$ is then passed to the pre-trained model.

Given a pre-trained backbone with frozen parameters, visual prompt tuning optimizes $\mathbf{p}$ to minimize a task-specific loss function. The resulting optimized prompt $\mathbf{p}^*$ enhances the backbone's performance on the target downstream task while keeping the backbone parameters unchanged.

Problem Statement

Machine unlearning requires that the influence of designated training data be completely removed from a trained model. When this removal must be provably equivalent to retraining from scratch on the reduced dataset, the requirement is called *exact unlearning*. We first formalize this notion and then state the specific problem addressed by LMEraser.

Definition 5.2 (*Exact Unlearning*) We adopt the notion of exact unlearning from Definition 1.18. Let T denote a fixed (possibly randomized) training procedure, let $\theta^* = T(D)$ be the released parameters, and let $D_f \subseteq D$ be a forget set. An unlearning operator U achieves exact unlearning if, for every measurable set $A \subseteq \Theta$,

$$\Pr\big[U(\theta^*, D, D_f) \in A\big] = \Pr\big[T(D \setminus D_f) \in A\big].$$

That is, the distribution of the unlearned model is identical to that of a model retrained from scratch on the retain set under the same training procedure.

Building on this definition, this chapter studies a large-scale vision model trained in two phases: a backbone M with parameters θ is first pre-trained on a public dataset D_p, and a collection of prompts and classifier heads is then tuned on a private dataset D_s, where

$$D = D_p \cup D_s, \qquad D_p \cap D_s = \varnothing.$$

Given a forget set $D_f \subseteq D_s$, the goal is to construct an unlearned model h_{θ_u} whose distribution matches retraining on the reduced dataset

$$D' = D \setminus D_f. \tag{5.2}$$

Design of Exact Unlearning for Large Models

LMEraser achieves efficient exact unlearning through three core designs that work in concert. First, a data partitioning strategy separates public and private data and freezes the backbone after pre-training, confining private data influence to a small set of learnable parameters. Second, an adaptive prompt tuning mechanism clusters private data by diversity and trains a dedicated prompt and classifier head for each cluster, enabling fine-grained feature recognition. Third, a targeted unlearning procedure removes the influence of designated data by re-tuning only the affected cluster's prompt and classifier head, with provable exact unlearning guarantees. The remainder of this section presents each design in detail.

Data Partitioning and Pre-training

Training large models on mixed public and private datasets entangles the influence of sensitive data throughout the model's parameters, creating privacy risks and making unlearning costly because each removal request necessitates retraining the entire model on the full training dataset [17]. LMEraser addresses this problem by partitioning the data at the source and restricting private data influence to a small, independently retrainable module.

Specifically, LMEraser partitions the training dataset $\mathcal{D}$ into a public dataset $\mathcal{D}_p$ and a private dataset $\mathcal{D}_s$ based on their provenance:

$$\mathcal{D} = \mathcal{D}_p \cup \mathcal{D}_s, \qquad \mathcal{D}_p \cap \mathcal{D}_s = \emptyset. \tag{5.3}$$

Public data $\mathcal{D}_p$, such as those from ImageNet [13] and COCO [14], are collected and cleansed by research institutions or non-profit organizations in compliance with legal and ethical standards. These data are widely accessible and have been extensively verified by the scientific community, ensuring their reliability and safety. Private data $\mathcal{D}_s$, such as medical records and bank transactions, contain sensitive information and are governed by strict privacy and security regulations.

LMEraser pre-trains a ViT backbone $\mathcal{M}$ using only the public dataset $\mathcal{D}_p$, as illustrated in Fig. 5.2. Denoting the distribution of $\mathcal{D}_p$ by $P_p(\mathcal{X})$, the pre-training objective is

$$\boldsymbol{\theta}^* = \arg\min_{\boldsymbol{\theta}} \ \mathbb{E}_{\mathbf{x}\sim P_p(\mathcal{X})}\big[\ell\big(\mathcal{M}_{\boldsymbol{\theta}}(\mathbf{x}),\ y\big)\big], \tag{5.4}$$

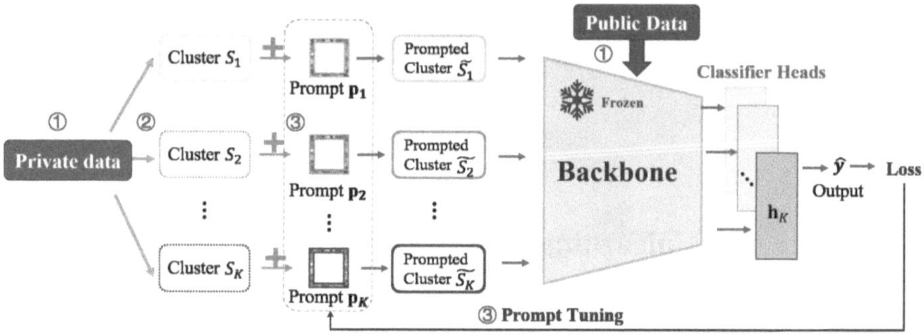

Fig. 5.2 Overview of the LMEraser training process: (1) Partition the training dataset into public and private datasets and pre-train the backbone on the public dataset; (2) Adaptively cluster the private data based on diversity; (3) Tune prompts and classifier heads for each cluster

where $\boldsymbol{\theta}$ denotes the backbone parameters, ℓ is the loss function, and y is the ground-truth label.

After pre-training, the backbone parameters $\boldsymbol{\theta}^*$ are frozen. Freezing achieves two objectives simultaneously: it isolates private data influence from the backbone so that unlearning private data never requires retraining the backbone, and it preserves the foundational knowledge acquired during pre-training, enhancing performance stability and preventing catastrophic forgetting in which removing a single data point could disrupt the knowledge learned from others [12].

Data partitioning and backbone freezing confine all private data influence to the prompt tuning stage. The next subsection describes how LMEraser adaptively clusters private data and trains dedicated prompts and classifier heads for each cluster, establishing the modular structure that enables efficient unlearning.

Adaptive Prompt Tuning

Although pre-training on public datasets captures broad foundational knowledge, it lacks domain-specific features that are essential for specialized tasks [18]. Private data, with their unique attributes and non-standard distributions, are valuable for bridging this gap through prompt tuning. However, the rich variability within private datasets poses significant challenges: a single prompt may suffice for low-diversity data, but is less effective for high-diversity data because it lacks the flexibility to capture complex variations [19, 20]. Furthermore, relying on a single prompt reduces unlearning efficiency, as removing any private data point necessitates retraining the prompt over the entire private dataset.

To address these challenges, LMEraser introduces an *adaptive prompt tuning mechanism* that clusters private data by diversity and then trains a dedicated prompt and classifier head for each cluster, as illustrated in Fig. 5.2. This subsection describes the three components of this mechanism: diversity-aware clustering, the underlying hierarchical clustering algorithm, and cluster-specific prompt optimization.

Algorithm 6 Adaptive Private Data Clustering

Require: Private dataset $\mathcal{D}_s$, pre-trained backbone $\mathcal{M}$
Ensure: Clustered private dataset $\{\mathbf{S}_1, \ldots, \mathbf{S}_K\}$ with cluster prototypes $\{\mathbf{o}_1, \ldots, \mathbf{o}_K\}$
$\mathcal{D}_{\text{sample}} \leftarrow \text{RandomSample}(\mathcal{D}_s)$
for each data point $\mathbf{x}_i$ in $\mathcal{D}_{\text{sample}}$ **do**
  $\mathbf{f}_i \leftarrow \mathcal{M}(\mathbf{x}_i)$ {Extract feature of $\mathbf{x}_i$}
end for
for each pair (i, j) in $\mathcal{D}_{\text{sample}}$ **do**
  $\mathbf{D}_{ij} \leftarrow \|\mathbf{f}_i - \mathbf{f}_j\|_2$ {Compute distance matrix}
end for
$\{\mathbf{S}_1, \ldots, \mathbf{S}_K\} \leftarrow \text{HierCluster}(\mathbf{D}, \mathcal{D}_{\text{sample}})$
for each cluster $\mathbf{S}_k$ **do**
  $\mathbf{o}_k \leftarrow \frac{1}{|\mathbf{S}_k|} \sum_{\mathbf{x}_i \in \mathbf{S}_k} \mathcal{M}(\mathbf{x}_i)$ {Prototype of $\mathbf{S}_k$}
end for
for each data point $\mathbf{x}_i \in \mathcal{D}_s \setminus \mathcal{D}_{\text{sample}}$ **do**
  $\mathbf{f}_i \leftarrow \mathcal{M}(\mathbf{x}_i)$ {Extract feature of $\mathbf{x}_i$}
  $k \leftarrow \arg\min_k \|\mathbf{f}_i - \mathbf{o}_k\|_2$
  Add $\mathbf{x}_i$ to cluster $\mathbf{S}_k$
end for
return $\{\mathbf{S}_1, \ldots, \mathbf{S}_K\}$, $\{\mathbf{o}_1, \ldots, \mathbf{o}_K\}$

Adaptive Private Data Clustering

Unlike traditional sharding-based unlearning strategies that randomly partition data into a predetermined number of shards without considering data characteristics, LMEraser adaptively clusters private data using Hierarchical Agglomerative Clustering (HierCluster) [21] and then trains tailored prompts and classifier heads for each cluster. A detailed description of the HierCluster algorithm is provided in section "HierCluster and Performance Analysis".

Algorithm 6 details the adaptive clustering process. Initially, a subset $\mathcal{D}_{\text{sample}}$ is sampled from the private dataset $\mathcal{D}_s$. For each data point $\mathbf{x}_i \in \mathcal{D}_{\text{sample}}$, features $\mathbf{f}_i = \mathcal{M}(\mathbf{x}_i)$ are extracted using the pre-trained backbone $\mathcal{M}$. A pairwise distance matrix $\mathbf{D}$ is computed from these features, and HierCluster partitions the sampled data into K clusters $\mathbf{S}_1, \ldots, \mathbf{S}_K$ (section "HierCluster and Performance Analysis"). For each cluster $\mathbf{S}_k$, a prototype $\mathbf{o}_k$ is computed as the mean feature vector of its members:

$$\mathbf{o}_k = \frac{1}{|\mathbf{S}_k|} \sum_{\mathbf{x}_i \in \mathbf{S}_k} \mathcal{M}(\mathbf{x}_i). \tag{5.5}$$

The remaining data points $\mathbf{x}_i \in \mathcal{D}_s \setminus \mathcal{D}_{\text{sample}}$ are then assigned to the cluster with the nearest prototype:

$$k_i^* = \arg\min_k \|\mathbf{f}_i - \mathbf{o}_k\|_2. \tag{5.6}$$

This sampling-based approach reduces clustering costs and enhances unlearning efficiency, because re-clustering is only required when unlearning involves data that was part of the initial sample $\mathcal{D}_{\text{sample}}$. By grouping similar data together, cluster-specific prompts can capture nuanced differences within each group, significantly enhancing the model's feature recognition and pattern analysis capabilities. Moreover, unlearning requests only require re-optimization of the affected cluster's prompt and classifier head rather than the entire private dataset.

HierCluster and Performance Analysis

Hierarchical Agglomerative Clustering (HierCluster) is an adaptive algorithm well suited to large-scale data applications. Unlike methods such as k-means that require a predefined number of clusters, HierCluster dynamically determines the cluster count based on the data's inherent structure and supports the discovery of clusters with arbitrary shapes [21].

Algorithm 7 HierCluster

Require: Dataset $\mathcal{D}$, distance matrix $\mathbf{D}$, distance threshold t
Ensure: Clusters $\{\mathbf{S}_1, \mathbf{S}_2, \ldots, \mathbf{S}_K\}$
Initialize each $\mathbf{x}_i \in \mathcal{D}$ as a singleton cluster $\mathbf{S}_i \leftarrow \{\mathbf{x}_i\}$
$\mathbb{S} \leftarrow \{\mathbf{S}_1, \mathbf{S}_2, \ldots, \mathbf{S}_n\}$ {Set of all clusters}
while true do
  $minDist \leftarrow \infty$
  $toMerge \leftarrow (null,\ null)$
  for each cluster pair $(\mathbf{S}_i, \mathbf{S}_j)$ in $\mathbb{S}$ **do**
    $dist \leftarrow$ average linkage distance between $\mathbf{S}_i$ and $\mathbf{S}_j$ using $\mathbf{D}$
    if $dist < minDist$ **then**
      $minDist \leftarrow dist$
      $toMerge \leftarrow (\mathbf{S}_i, \mathbf{S}_j)$
    end if
  end for
  if $minDist > t$ **or** $|\mathbb{S}| = 1$ **then**
    break
  end if
  $\mathbf{S}_{\text{new}} \leftarrow$ merge clusters in $toMerge$
  Update $\mathbb{S}$ by removing merged clusters and adding $\mathbf{S}_{\text{new}}$
end while
return $\{\mathbf{S}_1, \mathbf{S}_2, \ldots, \mathbf{S}_K\}$

Algorithm 7 details the iterative merging mechanism. The process begins by treating each data point $\mathbf{x}_i$ as a singleton cluster $\mathbf{S}_i$. At each iteration, the algorithm identifies the pair of clusters with the smallest average linkage distance (computed from the distance matrix $\mathbf{D}$) and merges them. Merging continues until the minimum inter-cluster distance exceeds a threshold t or all data points have been merged into a single cluster. Formally, the merging criterion is

$$\text{Merge}(\mathbf{S}_i,\ \mathbf{S}_j) \quad \text{if} \quad d(\mathbf{S}_i,\ \mathbf{S}_j) < t, \tag{5.7}$$

where $d(\mathbf{S}_i, \mathbf{S}_j)$ denotes the average linkage distance between clusters $\mathbf{S}_i$ and $\mathbf{S}_j$. The threshold t is set adaptively according to the desired granularity, ensuring that the algorithm captures the dataset's diversity without requiring a predefined cluster count. This adaptability makes HierCluster particularly suitable for the diverse nature of private datasets encountered in practice.

The efficiency and quality of HierCluster have been verified empirically. Under our experimental setup, clustering a sample of 1,000 data points from a private dataset was consistently completed in less than 3 s, representing approximately 0.1% of the total time required for 50 epochs of model training and making the clustering overhead negligible.

The clustering quality is further assessed through visual comparison. Figure 5.3a shows that HierCluster forms homogeneous clusters with clear inter-cluster separation, enabling LMEraser to learn nuanced features within each group and thereby enhancing classification accuracy. In contrast, Fig. 5.3b demonstrates that random partitioning produces groups that lack semantic coherence and clear separation. This contrast underscores the advantage of diversity-aware clustering for both model performance and unlearning efficiency.

Prompt Tuning for Clustered Data

After clustering, each cluster $\mathbf{S}_k$ is assigned a randomly initialized pixel-level border prompt $\mathbf{p}_k$. These prompts are then tuned using the private data within each cluster to optimize task performance.

Each prompt serves as a learnable transformation applied to the private data:

$$\tilde{\mathbf{x}}_i = g(\mathbf{x}_i,\ \mathbf{p}_k) = \mathbf{x}_i + \mathbf{p}_k, \tag{5.8}$$

where $\mathbf{x}_i$ is the original private data point assigned to cluster k_i^*, $\mathbf{p}_k$ is the prompt for that cluster, and $\tilde{\mathbf{x}}_i$ is the prompted input. As illustrated in Fig. 5.2, these prompts are applied as border overlays to the private images.

The prompt tuning process jointly optimizes the prompt $\mathbf{p}_k$ and the classifier head $\mathbf{h}_k$ for each cluster by minimizing the cross-entropy loss:

$$\mathbf{p}_k^*,\ \mathbf{h}_k^* = \arg\min_{\mathbf{p}_k,\,\mathbf{h}_k} \sum_{(\mathbf{x}_i, y_i)\in\mathbf{S}_k} \ell\big(H\big(\mathcal{M}(\mathbf{x}_i + \mathbf{p}_k;\ \boldsymbol{\theta}^*);\ \mathbf{h}_k\big),\ y_i\big), \tag{5.9}$$

where $\mathcal{M}(\cdot;\ \boldsymbol{\theta}^*)$ is the frozen pre-trained backbone, $H(\cdot;\ \mathbf{h}_k)$ is the classification function parameterized by the classifier head $\mathbf{h}_k$, and ℓ is the cross-entropy loss. Through iterative gradient descent, the prompt $\mathbf{p}_k$ adjusts the input distribution of cluster $\mathbf{S}_k$ to better align with the backbone's learned representations, while each classifier head specializes in its cluster's label space.

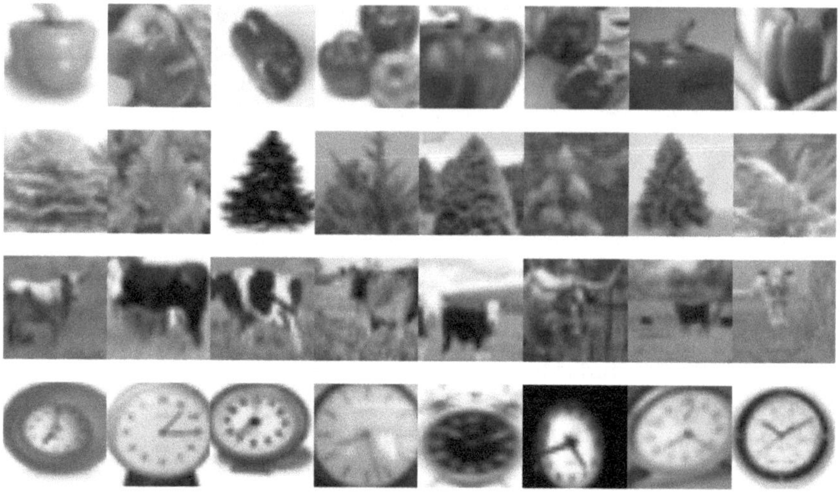

(a) Visualization of HierCluster results on CIFAR100. Each row represents a different cluster, showcasing images grouped by their visual similarity.

(b) Visualization of random partitioning on CIFAR100. Each row displays images from one group, illustrating the lack of semantic coherence.

Fig. 5.3 Comparison between HierCluster (top) and random partitioning (bottom) on the CIFAR100 dataset

Inference

During inference, given a test input $\mathbf{x}$, the system first identifies the nearest cluster $k^* = \arg\min_k \|\mathcal{M}(\mathbf{x}) - \mathbf{o}_k\|_2$, applies the corresponding optimized prompt $\mathbf{p}_{k^*}^*$ to obtain the prompted input $\tilde{\mathbf{x}} = g(\mathbf{x}, \mathbf{p}_{k^*}^*)$, and feeds $\tilde{\mathbf{x}}$ into the backbone and the cluster's classifier head $\mathbf{h}_{k^*}^*$ to produce the final prediction.

The adaptive prompt tuning mechanism partitions private data into diversity-aware clusters, each equipped with a dedicated prompt and classifier head. This modular design confines each private data point's influence to a single cluster and a small number of learnable parameters. The next subsection shows how this structure enables efficient and provably exact unlearning through targeted prompt re-tuning.

Prompt-Tuning-Based Unlearning

The modular prompt architecture established in the preceding subsection naturally supports efficient exact unlearning: because each private data point influences only its cluster's prompt and classifier head, removing that data point requires re-tuning only the affected module rather than the entire model. This subsection formalizes the unlearning process and establishes the theoretical properties that guarantee exactness.

The Unlearning Process

When an unlearning request targets a data point $(\mathbf{x}_r, y_r) \in \mathcal{D}_f$ that belongs to cluster $\mathbf{S}_k$, LMEraser re-tunes only the prompt $\mathbf{p}_k$ and classifier head $\mathbf{h}_k$ of the affected cluster. Formally, the cluster is first updated to $\mathbf{S}_k' = \mathbf{S}_k \setminus \{(\mathbf{x}_r, y_r)\}$ by excluding the designated data point. The prompt and classifier head are then re-optimized on the reduced cluster:

$$\mathbf{p}_k^{*\prime},\ \mathbf{h}_k^{*\prime} = \arg\min_{\mathbf{p}_k,\, \mathbf{h}_k} \sum_{(\mathbf{x}_i, y_i) \in \mathbf{S}_k'} \ell\big(H\big(\mathcal{M}(\mathbf{x}_i + \mathbf{p}_k;\ \boldsymbol{\theta}^*);\ \mathbf{h}_k\big),\ y_i\big), \tag{5.10}$$

where $\mathcal{M}(\cdot;\ \boldsymbol{\theta}^*)$ is the frozen backbone, $H(\cdot;\ \mathbf{h}_k)$ is the classification function, and ℓ is the loss function.

This targeted re-tuning strategy offers three significant advantages. First, cluster-specific prompt optimization isolates private data influence within its cluster, eliminating the need to trace changes across the entire model. Second, optimizing only the small-size prompt and classifier head (rather than the full backbone) significantly reduces computational cost and enhances unlearning efficiency. Third, combining pixel-level border prompts with a frozen pre-trained backbone preserves the backbone's learned representations, effectively minimizing performance degradation and the risk of catastrophic forgetting.

Exact Unlearning Guarantees

LMEraser provides exact unlearning guarantees at the algorithmic level. An alternative empirical approach to verifying unlearning is Membership Inference Attacks (MIA), but MIA faces inherent limitations that prevent it from serving as a reliable verification metric. When duplicate data points exist in the training set, successful unlearning of one copy still triggers MIA detection due to the other copy's legitimate presence, producing false positives. Conversely, forging methods [22] can manipulate a model to evade MIA detection even when the target data's influence remains, producing false negatives. For these reasons, LMEraser focuses on rigorous algorithmic-level guarantees rather than empirical verification.

The exactness of LMEraser rests on three key properties that follow directly from its architectural design.

Property 5.1 (Data Influence Isolation) *The backbone-prompt architecture enforces a strict separation of data influence. By Eq.* (5.3)*, the training data are partitioned as* $\mathcal{D} = \mathcal{D}_p \cup \mathcal{D}_s$ *with* $\mathcal{D}_p \cap \mathcal{D}_s = \emptyset$*. Public data* $\mathcal{D}_p$ *influence only the backbone parameters* $\boldsymbol{\theta}^*$*, while private data* $\mathcal{D}_s$ *affect only the prompts* $\{\mathbf{p}_k\}$ *and classifier heads* $\{\mathbf{h}_k\}$*. Because the backbone is frozen after pre-training, unlearning operations on private data cannot affect the backbone's learned representations.*

Property 5.2 (Cluster Independence) *The adaptive clustering mechanism ensures that each cluster's prompt and classifier head are optimized independently (Eq.* (5.9)*). For any two distinct clusters* i *and* j *with their respective parameters* $(\mathbf{p}_i, \mathbf{h}_i)$ *and* $(\mathbf{p}_j, \mathbf{h}_j)$*, modifying cluster* i*'s prompt and classifier head has no impact on cluster* j*'s decision boundaries or feature representations.*

Property 5.3 (Complete Removal) *Complete retraining of the affected cluster's prompt and classifier head guarantees the removal of all direct and indirect influences of the unlearned data point. The re-optimized parameters* $\mathbf{p}_k^{*\prime}$, $\mathbf{h}_k^{*\prime}$ *(Eq.* (5.10)*) are derived solely from the reduced cluster* $\mathbf{S}_k'$*, which excludes the target data point entirely.*

These three properties collectively establish that LMEraser achieves exact unlearning at the algorithmic level. Data Influence Isolation (Property 5.1) confines private data influence to the prompt layer; Cluster Independence (Property 5.2) ensures that re-tuning one cluster does not affect others; and Complete Removal (Property 5.3) guarantees that the re-optimized parameters carry no residual influence from the removed data. Together, these properties provide stronger guarantees than empirical verification methods.

Theoretical Analysis

In practical deployments, model owners often do not perform unlearning immediately upon receiving each individual request. Instead, they collect requests over a batch period and execute them together, optimizing resource usage at the cost of some delay. This section develops a theoretical framework to quantify LMEraser's efficiency in such batch unlearning scenarios. We first model unlearning requests as a Poisson process (section "Probabilistic Model for Unlearning Requests"), then analyze the expected number of affected clusters (section "Analysis of Affected Clusters"), and finally derive the expected processing time and the efficiency gain of LMEraser over single-prompt methods (section "Processing Time Analysis and Efficiency Gains").

Probabilistic Model for Unlearning Requests

Consider a system with K clusters, where cluster i contains m_i data points for $i \in \{1, \ldots, K\}$, and the total number of private data points is $N = \sum_{i=1}^{K} m_i$. We model unlearning requests as a Poisson process with a total arrival rate λ for the entire dataset, and we assume that the probability of a request targeting any specific data point is uniform across all data points. The batch collection period is denoted by T.

Under the uniform targeting assumption, the rate at which cluster i receives unlearning requests is proportional to its size:

$$\lambda_i = \lambda \cdot \frac{m_i}{N}. \tag{5.11}$$

Because the Poisson process is memoryless and each data point is targeted independently, the number of unlearning requests arriving at cluster i during the interval $[0, T]$ follows a Poisson distribution with parameter $\lambda_i T$.

Analysis of Affected Clusters

A cluster is *affected* if it receives at least one unlearning request during the batch period $[0, T]$. In LMEraser, only affected clusters require re-tuning, so the number of affected clusters directly determines the scope of the unlearning operation. This subsection derives the expected value and variance of the number of affected clusters and establishes an upper bound that is independent of the cluster size distribution.

Expected Number of Affected Clusters

The probability that cluster i receives at least one unlearning request during $[0, T]$ is the complement of receiving none:

$$P_i = 1 - e^{-\lambda_i T} = 1 - e^{-\frac{m_i}{N}\lambda T}, \tag{5.12}$$

where λ_i is defined in Eq. (5.11). The expected number of affected clusters is therefore

$$\begin{aligned}\mathbb{E}[K_{\text{affected}}] &= \sum_{i=1}^{K} P_i = \sum_{i=1}^{K} \left(1 - e^{-\frac{m_i}{N}\lambda T}\right) \\ &\leq \sum_{i=1}^{K} \frac{m_i}{N}\lambda T = \lambda T,\end{aligned} \tag{5.13}$$

where the inequality follows from $1 - e^{-x} \leq x$ for all $x \geq 0$. This upper bound shows that the expected number of affected clusters is at most λT, regardless of how data points are distributed across clusters. When unlearning requests are infrequent (i.e., $\frac{m_i}{N}\lambda T$ is small for all i), the bound is tight, and the expected number of affected clusters depends only on the request rate λ and the batch period T.

Variance of the Number of Affected Clusters

The variance quantifies the spread around the expected value and is given by

$$\text{Var}[K_{\text{affected}}] = \sum_{i=1}^{K} P_i(1 - P_i). \tag{5.14}$$

Substituting the expression for P_i from Eq. (5.12) yields

$$\begin{aligned}\text{Var}[K_{\text{affected}}] &= \sum_{i=1}^{K} \left(1 - e^{-\frac{m_i}{N}\lambda T}\right) e^{-\frac{m_i}{N}\lambda T} \\ &\leq \sum_{i=1}^{K} \frac{m_i}{N}\lambda T \cdot 1 = \lambda T,\end{aligned} \tag{5.15}$$

where the inequality uses $e^{-x} \leq 1$ for $x \geq 0$. The variance is thus also bounded by λT. When unlearning requests are relatively infrequent, the variance remains small, indicating a tight concentration of the number of affected clusters around its expected value.

Processing Time Analysis and Efficiency Gains

The practical efficiency of an unlearning method is ultimately measured by its processing time. This subsection derives the expected processing time for both LMEraser and the single-prompt baseline, and then quantifies the efficiency gain. Because LMEraser's clusters are independent (Property 5.2), re-tuning operations for different affected clusters can be executed in parallel, so the total processing time depends on whether *any* cluster is affected rather than on *how many* are affected.

Expected Processing Time for LMEraser

Let T_c denote the time required to re-tune a single cluster's prompt and classifier head. The probability that at least one cluster is affected during $[0, T]$ is

$$\begin{aligned} P(\text{any cluster affected}) &= 1 - \prod_{i=1}^{K} e^{-\lambda_i T} \\ &= 1 - e^{-\sum_{i=1}^{K} \lambda_i T} = 1 - e^{-\lambda T}. \end{aligned} \tag{5.16}$$

Because all affected clusters can be re-tuned in parallel, the expected processing time is

$$\mathbb{E}[T_{\text{LMEraser}}] = T_c \cdot P(\text{any cluster affected}) = T_c \left(1 - e^{-\lambda T}\right). \tag{5.17}$$

For small λT (i.e., infrequent requests), using the first-order approximation $1 - e^{-x} \approx x$ gives

$$\mathbb{E}[T_{\text{LMEraser}}] \approx T_c \cdot \lambda T. \tag{5.18}$$

Expected Processing Time for Single-Prompt Method

In the single-prompt method, any unlearning request necessitates re-tuning the prompt over the entire private dataset. Let T_s denote the time required for this full re-tuning. The probability that at least one request arrives during $[0, T]$ is

$$P(\text{any unlearning}) = 1 - e^{-\lambda T}. \tag{5.19}$$

The expected processing time is therefore

$$\mathbb{E}[T_{\text{Single}}] = T_s \cdot P(\text{any unlearning}) = T_s \left(1 - e^{-\lambda T}\right). \tag{5.20}$$

For small λT, the approximation yields

$$\mathbb{E}[T_{\text{Single}}] \approx T_s \cdot \lambda T. \tag{5.21}$$

Time Efficiency Gain

The efficiency gain of LMEraser over the single-prompt method is defined as the ratio of their expected processing times:

$$\text{Efficiency Gain} = \frac{\mathbb{E}[T_{\text{Single}}]}{\mathbb{E}[T_{\text{LMEraser}}]} = \frac{T_s\left(1 - e^{-\lambda T}\right)}{T_c\left(1 - e^{-\lambda T}\right)} = \frac{T_s}{T_c}. \tag{5.22}$$

This result reveals that the efficiency gain is determined entirely by the ratio of the per-operation processing times and is independent of the request rate λ, the batch period T, and the cluster size distribution. Since re-tuning a single cluster's prompt and classifier head involves far fewer data points and parameters than re-tuning over the entire private dataset, $T_c \ll T_s$, and the efficiency gain is substantial. In our experiments (section "Experimental Evaluation"), LMEraser achieves approximately a 100-fold reduction in unlearning time compared to single-prompt methods, consistent with this theoretical prediction.

The theoretical analysis in this section has established three key results. First, the expected number of affected clusters is bounded by λT and is independent of the cluster size distribution (Eq. (5.13)). Second, the variance of the number of affected clusters shares the same bound, confirming predictable unlearning workloads (Eq. (5.15)). Third, the efficiency gain of LMEraser equals T_s/T_c, which depends solely on the ratio of per-operation costs and yields substantial speedups because cluster-level re-tuning involves far fewer parameters than full-dataset re-tuning (Eq. (5.22)). Beyond computational efficiency, LMEraser's parallel prompt architecture also enhances modularity, as individual prompts can be updated independently, and scalability, as new prompts can be incorporated seamlessly when data clusters evolve. The next section empirically validates these theoretical findings through comprehensive experiments.

Experimental Evaluation

LMEraser is evaluated along two complementary dimensions: *model utility*, measured by test accuracy on image classification tasks to verify that the model's core functionality remains intact after unlearning, and *unlearning efficiency*, measured by the time, computational, and storage costs required to process unlearning requests. Together, these two dimensions demonstrate that LMEraser can remove data influence completely without sacrificing classification performance or incurring prohibitive overhead.

Experimental Setup

This subsection describes the datasets, backbones, baselines, and implementation details used throughout the evaluation.

Private Datasets

We employ five diverse datasets as private datasets: HAM10000 [23], CIFAR10 [24], CIFAR100 [24], GTSRB [25], and SVHN [26]. These datasets are selected for their varying image sizes, number of images, class distributions, and degrees of diversity, providing a robust evaluation across different levels of task complexity. Details of the private datasets are summarized in Table 5.1.

Public Datasets

We utilize ImageNet-1K and ImageNet-22K as public datasets. These datasets provide a comprehensive foundation for the model backbones, ensuring a broad representation of general image features. Detailed specifications are summarized in Table 5.2.

Pre-trained Backbones

We primarily use ViT-B-22K as the backbone, pre-trained on ImageNet-22K with 86M parameters. We also employ ViT-B-1K [15] and Swin-B-1K (88M parameters) [27], both pre-trained on ImageNet-1K, as well as Swin-B-22K, pre-trained on ImageNet-22K [13].

Table 5.1 Statistical details of private datasets

Dataset	Image dimension	# Images	# Classes
HAM10000	$600 \times 450 \times 3$	10,015	7
CIFAR10	$32 \times 32 \times 3$	60,000	10
CIFAR100	$32 \times 32 \times 3$	60,000	100
SVHN	$32 \times 32 \times 3$	99,289	10
GTSRB	From $15 \times 15 \times 3$ to $250 \times 250 \times 3$	39,270	43

Table 5.2 Statistical details of public datasets

Dataset	Image dimension	# Images	# Classes
ImageNet-1K	$224 \times 224 \times 3$	1.28M	1,000
ImageNet-22K	$224 \times 224 \times 3$	14M	22,000

Baselines

We compare LMEraser against four methods. *Retraining* retrains the entire large model from scratch for each unlearning request, serving as the gold-standard reference. *SISA* [8] partitions training data into shards, trains a separate model on each shard, and produces final predictions through majority voting; unlearning requires retraining only the affected shard model. *Single-Prompt Unlearning*, adapted from Visual Prompting (VP) [19], uses a single prompt and a single classifier head for private data with a frozen backbone pre-trained on public data, retraining the prompt and classifier head for each unlearning request. *DAM-VP Unlearning*, based on DAM-VP [20], clusters private data by diversity and generates a prompt for each cluster without a classifier head, retraining the affected prompt for each unlearning request. Due to the scarcity of exact unlearning methods designed specifically for large models, the latter two adaptations help assess whether existing prompt tuning methods can be effectively repurposed for unlearning.

Implementation Details

Experiments were conducted on eight Nvidia Tesla V100-FHHL GPUs, each with 16 GB memory, complemented by an Intel Xeon CPU E5-2650 v4 at 2.20 GHz with 48 cores and 251 GB of RAM. The software environment consisted of Ubuntu 22.04.3 LTS (64-bit), PyTorch v2.1.2, CUDA 12.1, and Python 3.11.5. Our code is available at: https://github.com/lmeraser/lmeraser.

Private datasets were clustered using hierarchical agglomerative clustering (section "HierCluster and Performance Analysis"). The default configuration sampled 1,000 data points from each private dataset with a distance threshold of 10, yielding varying numbers of clusters across datasets: 194 for CIFAR10, 347 for CIFAR100, 30 for SVHN, and 90 for GTSRB. The number of clusters corresponds to the number of prompts used per dataset. Private images were standardized by resizing to 256×256 pixels and then center-cropping to 224×224 pixels. A 30-pixel border prompt was applied to each image, consistent with the method used in DAM-VP [20].

Evaluation of Model Utility

A practical unlearning method must preserve the model's predictive performance after data removal. This subsection evaluates LMEraser's model utility from four perspectives: classification accuracy relative to baselines, robustness across backbone architectures, resilience to increasing data removal ratios, and sensitivity to the number of prompts.

Comparing Classification Accuracy

To validate the utility of LMEraser, we compared its test accuracy against baseline methods using ViT-B-22K as the backbone. Figure 5.4 shows that LMEraser achieves comparable performance to the single-prompt method while outperforming DAM-VP. DAM-VP adopts a head-missing approach based on activation-feature to label mapping, which falls short in tasks requiring precise classification. In contrast, LMEraser's use of cluster-specific prompts and classifier heads allows for more nuanced feature recognition, resulting in superior performance. The single-prompt method, while effective in terms of accuracy, requires substantially longer time to handle each unlearning request (Table 5.3). LMEraser thus strikes a better balance between accuracy and computational efficiency.

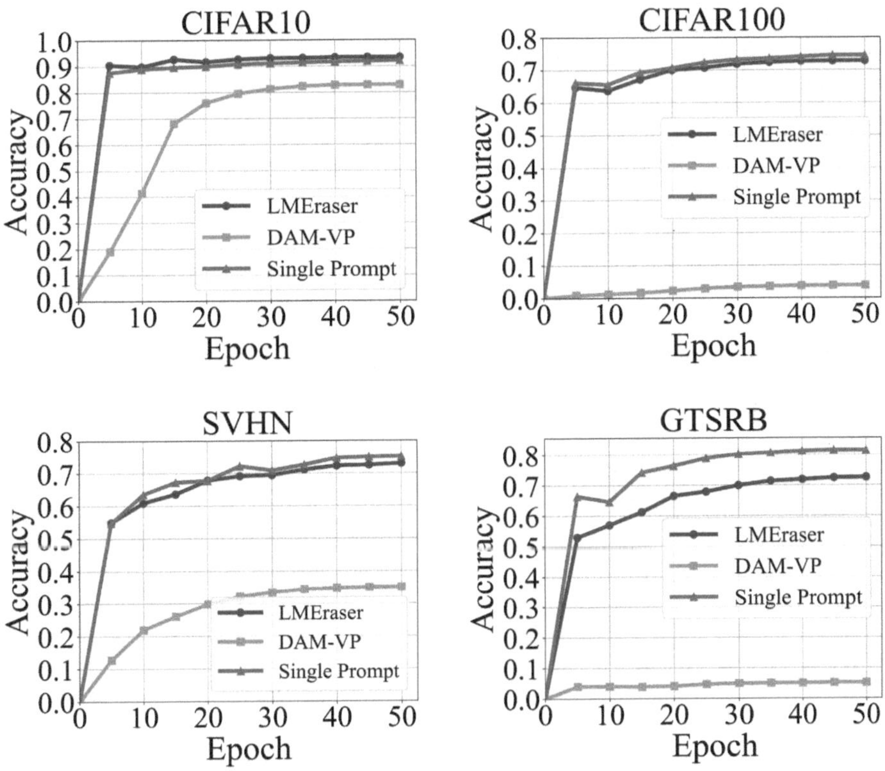

Fig. 5.4 Classification accuracy comparison between LMEraser and baseline methods on ViT-B-22K

Robustness across Different Backbones

To evaluate LMEraser's adaptability, we assessed its test accuracy using various backbone architectures while maintaining consistent conditions and prompt numbers for each dataset. Figure 5.5 demonstrates that LMEraser achieves high accuracy across diverse datasets and backbones, highlighting its versatility. Models pre-trained on ImageNet-22K (Swin-B-22K and ViT-B-22K) consistently outperform their ImageNet-1K counterparts, underscoring the importance of pre-training data scale and quality.

To further verify LMEraser's effectiveness on high-resolution domain-specific tasks, we conducted experiments on the HAM10000 medical dataset, which contains 10,015 dermatoscopic images (600 × 450 pixels) across seven categories of skin lesions. Figure 5.6 shows that LMEraser maintains strong performance on this challenging medical domain, achieving converged accuracy of 0.7910 (ViT-B-1K), 0.7946 (ViT-B-22K), 0.7872 (Swin-B-1K), and 0.8142 (Swin-B-22K). These results demonstrate that LMEraser can effectively handle both standard vision benchmarks and high-resolution domain-specific applications.

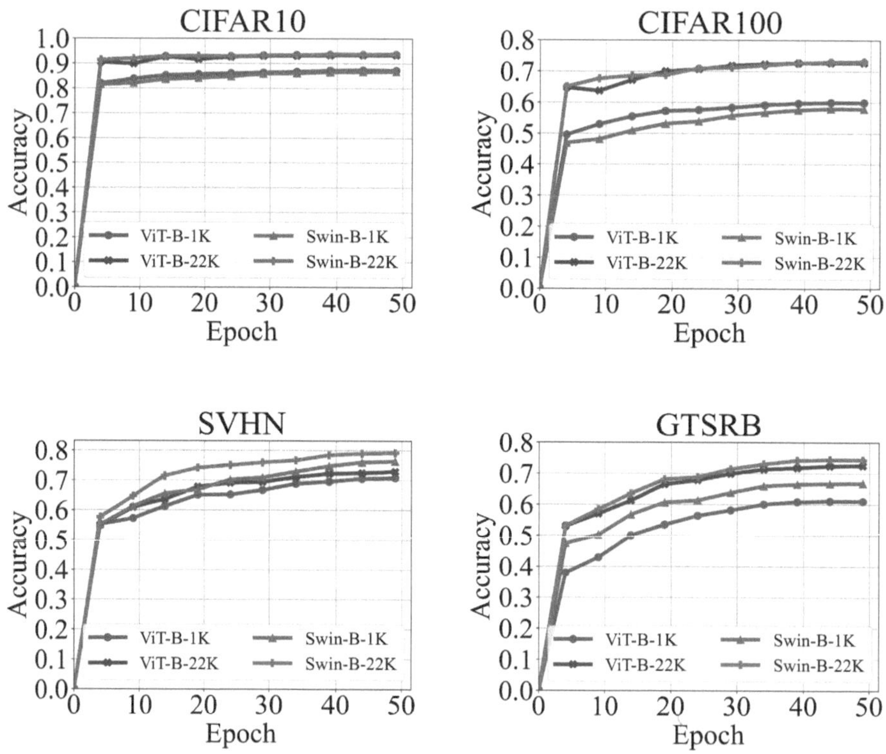

Fig. 5.5 Test accuracy of LMEraser using different backbones

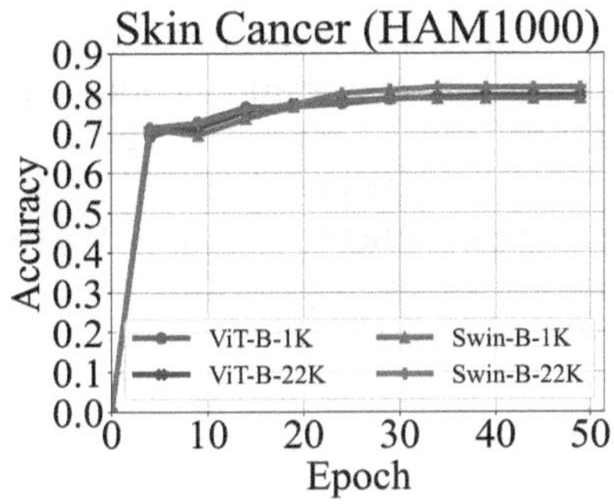

Fig. 5.6 Test accuracy of LMEraser on the HAM10000 medical dataset using different backbones

Resilience to Data Removal

We evaluated the impact of data removal on model stability by varying the unlearning ratio, defined as the percentage of private data points requested for removal, from 0% to 50% in 10% increments. At each increment, we randomly selected and removed the corresponding percentage of data points from the private dataset, then retrained the affected prompts and tested the accuracy. Figure 5.7 (Left) shows a marginal decrease in test accuracy from 0.7275 to 0.6953 (a 4.43% reduction) despite removing 50% of private data, demonstrating LMEraser's resilience to substantial data removal.

Impact of Prompt Quantity

We analyzed the relationship between prompt quantity and model performance by adjusting the clustering distance threshold from 7 to 11, which varies the number of prompts accordingly. Figure 5.7 (Right) shows that increasing the number of prompts

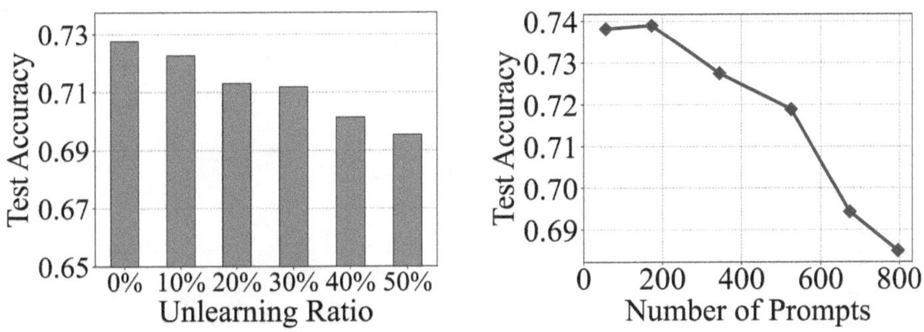

Fig. 5.7 LMEraser's test accuracy with various unlearning ratios (Left) and number of prompts (Right), using the ViT-B-22K backbone on the CIFAR100 private dataset

leads to a moderate decrease in test accuracy, because a larger number of prompts may reduce generalization and lead to overfitting within individual clusters.

Across all five private datasets and four backbone architectures, LMEraser achieves classification accuracy comparable to the single-prompt method and superior to DAM-VP, while remaining resilient to data removal ratios up to 50%. The next subsection evaluates the efficiency gains that LMEraser provides when processing unlearning requests.

Evaluation of Unlearning Efficiency

An exact unlearning method must not only preserve model utility but also process removal requests within practical time and resource budgets. This subsection evaluates LMEraser's unlearning efficiency along three metrics: the amount of affected training data, the number of affected model parameters, and the wall-clock time per unlearning request. We also present a theoretical analysis of batch unlearning performance.

Comparing Affected Training Data and Model Size

To assess resource efficiency, we compared the number of affected training data points and affected model parameters per unlearning request across methods, using CIFAR10 with the number of clusters fixed at ten for all datasets. Figure 5.8 (Left) shows that LMEraser and other prompt-tuning-based methods significantly reduce the amount of affected training data; LMEraser affects approximately 100 times fewer

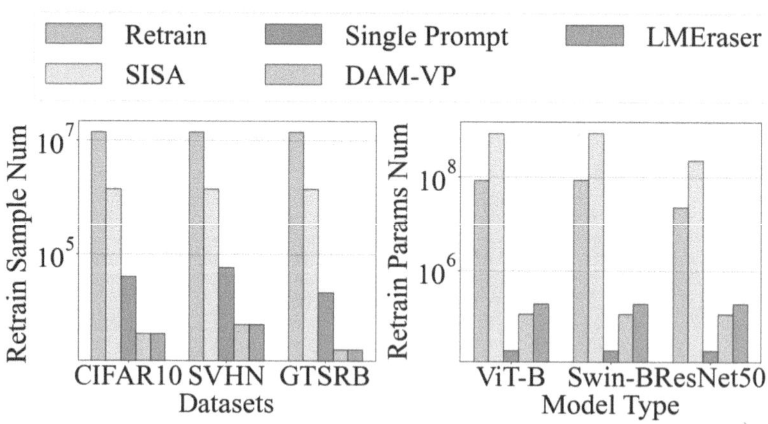

Fig. 5.8 Comparative analysis of LMEraser and baseline methods in terms of affected training data (Left) and model parameters (Right) for a single unlearning request

training data points compared to retraining and SISA. Figure 5.8 (Right) shows an even more dramatic improvement in affected model parameters, where LMEraser affects approximately 1,000 times fewer parameters compared to retraining and SISA.

Time Efficiency in Unlearning

We compare the average wall-clock time required to handle a single unlearning request across all methods, as shown in Table 5.3. Traditional methods, such as retraining and SISA, require days to process a single request on large backbones. In contrast, prompt-based methods such as LMEraser and DAM-VP reduce the processing time from days to seconds or minutes. This dramatic reduction is enabled by the parallel prompt architecture, which confines each unlearning operation to a small cluster rather than the entire model.

Theoretical Analysis of Batch Unlearning

In practice, model owners often collect unlearning requests over a period and process them in batches rather than individually. We develop a theoretical framework for analyzing LMEraser's performance in such batch scenarios, with full derivations presented in section "Theoretical Analysis". The framework models unlearning requests as a Poisson process with rate λ over a batch collection period T. Consider a model with K clusters, each containing m_i data points ($i \in \{1, \ldots, K\}$), and a total of $N = \sum_{i=1}^{K} m_i$ data points. The expected number of affected clusters in the interval $[0, T]$ is

$$\mathbb{E}[K_{\text{affected}}] = \sum_{i=1}^{K} \left(1 - e^{-\frac{m_i}{N}\lambda T}\right). \tag{5.23}$$

For small values of $\frac{m_i}{N}\lambda T$ (i.e., when unlearning requests are infrequent), this expression simplifies to

$$\mathbb{E}[K_{\text{affected}}] \approx \lambda T, \tag{5.24}$$

Table 5.3 Average time per unlearning request with the ViT-B-22K backbone across different datasets and methods

Dataset	Retrain	SISA	SinglePrompt (s)	DAM-VP (s)	LMEraser (s)
CIFAR10	Days	Days	3185	21.26	27.61
CIFAR100	Days	Days	4048	12.14	12.45
SVHN	Days	Days	4688	260.07	213.81
GTSRB	Days	Days	1740	26.64	27.73

revealing that the expected number of affected clusters becomes independent of cluster size distribution and depends solely on the request rate λ and the time interval T.

The expected processing time for LMEraser is $\mathbb{E}[T_{\text{LMEraser}}] = T_c(1 - e^{-\lambda T})$, where T_c is the time required to re-tune a single cluster's prompt and classifier head. For the single-prompt method, the expected processing time is $\mathbb{E}[T_{\text{Single}}] = T_s(1 - e^{-\lambda T})$, where T_s is the time to re-tune over the entire private dataset. The efficiency gain of LMEraser over the single-prompt method is therefore

$$\text{Efficiency Gain} = \frac{\mathbb{E}[T_{\text{Single}}]}{\mathbb{E}[T_{\text{LMEraser}}]} = \frac{T_s}{T_c}. \tag{5.25}$$

Because re-tuning a single cluster involves substantially less data than re-tuning over the entire private dataset, $T_c \ll T_s$, and LMEraser achieves substantial efficiency gains in batch unlearning scenarios.

LMEraser reduces affected training data by two orders of magnitude and affected model parameters by three orders of magnitude relative to retraining and SISA, while processing each unlearning request in seconds rather than days. The theoretical analysis confirms that these efficiency advantages extend to batch unlearning scenarios. The next subsection presents ablation studies that isolate the contributions of LMEraser's individual design components.

Ablation Studies

The preceding subsections have demonstrated LMEraser's overall effectiveness in preserving model utility and achieving efficient unlearning. This subsection isolates the contributions of two key design choices through ablation experiments: the adaptive clustering mechanism and the use of multiple classifier heads.

Adaptive Prompt Tuning

To evaluate the effectiveness of adaptive clustering, we compared LMEraser with a random partitioning approach (RndPart Prompt) in which private data is randomly partitioned before prompt tuning, while all other conditions remain constant. Figure 5.9 shows that both methods achieve similar unlearning times, but RndPart Prompt yields significantly lower test accuracy, especially on complex tasks such as CIFAR100. This result confirms that LMEraser's adaptive clustering captures inherent data characteristics that are essential for accurate prompt tuning.

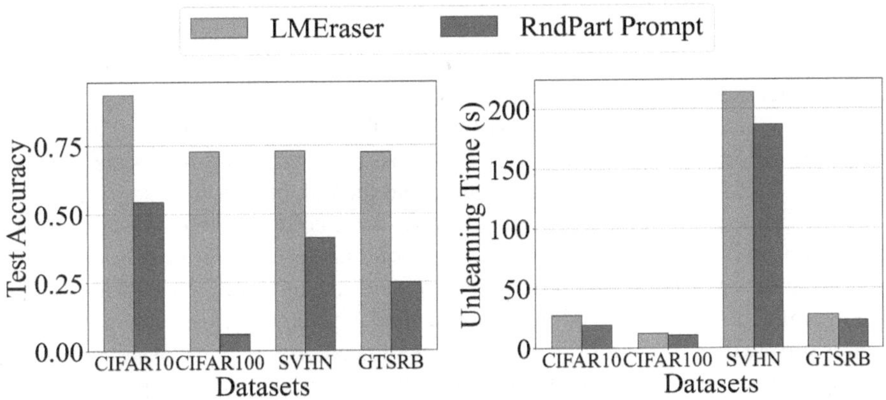

Fig. 5.9 Performance comparison between random partitioning and LMEraser's adaptive clustering across different datasets

Multiple Classifier Heads

To evaluate the effectiveness of multiple classifier heads, we compared LMEraser with a single-classifier-head variant while all other conditions remained constant. Figure 5.10 shows that both methods achieve similar test accuracy across datasets, but LMEraser requires substantially less unlearning time. The single-classifier-head method must retrain the shared head for every unlearning request, consuming more time and resources. In contrast, LMEraser's multiple classifier heads enable targeted retraining of only the affected head, significantly improving unlearning efficiency.

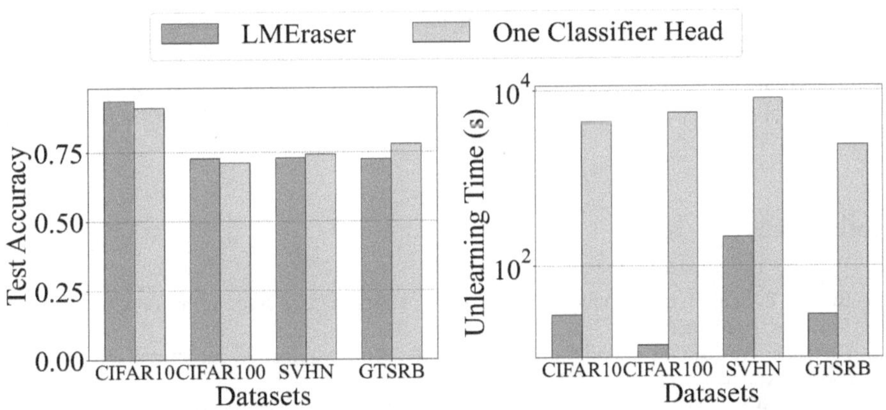

Fig. 5.10 Performance comparison between a single classifier head and LMEraser's multiple classifier heads across different datasets

Conclusion

This chapter has presented LMEraser, an efficient and exact unlearning method specifically designed for large vision models. The design of LMEraser addresses the three challenges: tracing data influence across billions of parameters, the computational cost of modifying such parameters, and the risk of catastrophic forgetting.

LMEraser tackles these challenges through three complementary designs. First, partitioning the training data into public and private subsets and freezing the backbone after pre-training on public data confines all private data influence to the prompt layer, making influence tracing straightforward. Second, an adaptive clustering mechanism groups private data by diversity and trains a dedicated prompt and classifier head for each cluster, reducing the computational scope of each unlearning operation from the entire model to a single small module. Third, the frozen backbone preserves foundational knowledge throughout the unlearning process, effectively eliminating the risk of catastrophic forgetting. Together, these designs guarantee exact unlearning at the algorithmic level through three provable properties: data influence isolation, cluster independence, and complete removal.

Comprehensive experiments across five private datasets, four backbone architectures, and four baseline methods confirm that LMEraser achieves classification accuracy comparable to the single-prompt method and superior to DAM-VP, while reducing unlearning time from days to seconds, affected training data by two orders of magnitude, and affected model parameters by three orders of magnitude relative to retraining and SISA. Ablation studies further validate that both adaptive clustering and multiple classifier heads contribute meaningfully to this performance.

One limitation of LMEraser is that it assumes a clear separation between public and private data at the source level, which may not hold in all application scenarios. In addition, the current design targets vision models with a pre-train-then-tune paradigm; extending the approach to other modalities and training regimes remains an open direction. The next chapter addresses a complementary challenge: enabling unlearning during the training process itself, eliminating the need for post-hoc removal by integrating unlearning directly into the model's learning dynamics.

References

1. Brown, T., et al.: Language models are few-shot learners. Adv. Neural. Inf. Process. Syst. **33**, 1877–1901 (2020)
2. Dehghani, M.: Scaling vision transformers to 22 billion parameters. In: International Conference on Machine Learning. PMLR, pp. 7480–7512 (2023)
3. Radford, A.: Learning transferable visual models from natural language supervision. In: International Conference on Machine Learning. PMLR, pp. 8748–8763 (2021)
4. Wu, Y.: Quantifying privacy risks of prompts in visual prompt learning. arXiv preprint arXiv:2310.11970 (2023)
5. Voigt, P., Von dem Bussche, A.: The EU General Data Protection Regulation (GDPR). A Practical Guide. Springer International Publishing (2017)

6. State of California Department of Justice Office of the Attorney General. California Consumer Privacy Act (CCPA). https://oag.ca.gov/privacy/ccpa (2023)
7. Ginart, A.A., et al.: Making AI forget you: data deletion in machine learning. In: Advances in Neural Information Processing Systems, vol. 32, pp. 1–14. NeurIPS (2019). eprint: 1907.05012
8. Bourtoule, L.: Machine unlearning. In: IEEE Symposium on Security and Privacy (SP), pp. 141–159. IEEE (2021)
9. Guo, C.: Certified data removal from machine learning models. In: International Conference on Machine Learning. PMLR, pp. 3832–3842 (2020)
10. Suriyakumar, V., Wilson, A.C.: Algorithms that approximate data removal: new results and limitations. In: Advances in Neural Information Processing Systems, vol. 35, pp. 18892–18903 (2022)
11. Wu, G., Hashemi, M., Srinivasa, C.: Puma: performance unchanged model augmentation for training data removal. In: Proceedings of the AAAI Conference on Artificial Intelligence, vol. 36(8), pp. 8675–8682 (2022)
12. Shibata, T., et al.: Learning with selective forgetting. In: IJCAI International Joint Conference on Artificial Intelligence, pp. 989–996 (2021). https://doi.org/10.24963/ijcai.2021/137. ISBN: 9780999241196
13. Deng, J.: ImageNet: a large-scale hierarchical image database. In: IEEE Conference on Computer Vision and Pattern Recognition, pp. 248–255. IEEE (2009). https://doi.org/10.1109/CVPR.2009.5206848
14. Lin, T.-Y.: Microsoft COCO: common objects in context. In: European Conference on Computer Vision, pp. 740–755 (2014)
15. Dosovitskiy, A., et al.: An image is worth 16 × 16 words: transformers for image recognition at scale. In: International Conference on Learning Representations (2020)
16. Vaswani, A.: Attention is all you need. In: Advances in Neural Information Processing Systems, vol. 30 (2017)
17. Golatkar, A.: Mixed-privacy forgetting in deep networks. In: Proceedings of the IEEE/CVF Conference on Computer Vision and Pattern Recognition, pp. 792–801 (2021)
18. Bommasani, R.: On the opportunities and risks of foundation models. arXiv preprint arXiv:2108.07258 (2021)
19. Bahng, H.: Exploring visual prompts for adapting large-scale models. arXiv preprint arXiv:2203.17274 (2022)
20. Huang, Q.: Diversity-aware meta visual prompting. In: Proceedings of the IEEE/CVF Conference on Computer Vision and Pattern Recognition, pp. 10878–10887 (2023)
21. Müllner, D.: Modern hierarchical, agglomerative clustering algorithms. arXiv preprint arXiv:1109.2378 (2011)
22. Thudi, A.: On the necessity of auditable algorithmic definitions for machine unlearning. In: 31st USENIX Security Symposium (USENIX Security 22), pp. 4007–4022 (2022)
23. Tschandl, P., Rosendahl, C., Kittler, H.: The HAM10000 dataset, a large collection of multi-source dermatoscopic images of common pigmented skin lesions. Sci. Data **5**(1), 1–9 (2018)
24. Krizhevsky, A., Hinton, G.: Learning multiple layers of features from tiny images. University of Toronto, Technical report (2009)
25. Stallkamp, J.: Man vs. computer: benchmarking machine learning algorithms for traffic sign recognition. Neural Netw. **32**, 323–332 (2012)
26. Netzer, Y., et al.: Reading digits in natural images with unsupervised feature learning. In: NIPS Workshop on Deep Learning and Unsupervised Feature Learning, vol. 2, p. 5 (2011)
27. Liu, Z.: Swin transformer: hierarchical vision transformer using shifted windows. In: Proceedings of the IEEE/CVF International Conference on Computer Vision, pp. 10012–10022 (2021)

Chapter 6
Approximate Unlearning for Large Models via Noise Injection

Abstract Existing unlearning methods operate after training completes, requiring either full retraining or post-hoc corrections at the converged state. For training runs that last days or weeks, neither halting to retrain nor waiting until convergence is attractive. This chapter develops a theoretical framework for *in-training* certified unlearning that exploits the natural attenuation of data influence in regularized optimization. We analyze a simplified but tractable setting: single-pass mini-batch SGD with coupled L_2 regularization. Building on the window-accumulated influence analysis of Chap. 4, we show that under a local contractivity condition on the optimization dynamics, the influence of a single data point decays geometrically through subsequent optimization steps. After a sufficient number of steps, the residual influence is small enough to be masked by calibrated Gaussian noise, yielding (ε, δ)-certified unlearning relative to a noisy retraining oracle, without explicit influence computation, gradient storage, or Hessian inversion. We provide theoretical guarantees on the decay rate, the required noise magnitude, and the resulting utility loss.

Motivation and Contribution

Training a large language model may consume thousands of GPU-hours over several weeks, processing hundreds of billions of tokens drawn from web crawls, books, code repositories, and curated corpora [1–3]. The scale of these datasets makes comprehensive pre-training curation impractical: despite extensive filtering pipelines, problematic data routinely enters the training set and may not be identified until training is well underway [4]. A content provider may issue a copyright takedown, a privacy audit may flag personally identifiable information, an adversarial analysis may reveal poisoned samples, or a regulatory body may mandate deletion under frameworks such as the GDPR [5]. In each case, the demand is the same: remove the influence of specific training data from the model, ideally with formal guarantees [6, 7].

J. Xu and X. Jia, *Machine Unlearning*, Wireless Networks,
https://doi.org/10.1007/978-3-032-30978-5_6

Existing unlearning methods largely treat this demand as a post-training problem [7–11]. Exact methods require full retraining; approximate methods apply corrections at the converged state [6, 7, 9–11]. Neither approach is satisfactory for long-running training jobs. This practical gap motivates a shift from post-training to *in-training* unlearning: the ability to accept deletion requests during training and execute them once the data point's residual influence has become small enough, without halting or restarting. The desired mechanism should satisfy four requirements simultaneously: certified removal guarantees, utility preservation, computational efficiency, and the ability to operate without stored gradient histories [7, 11].

This chapter takes a first step toward such a mechanism by analyzing a simplified but theoretically tractable setting. Chapter 4 introduced the propagator matrix $Z_t = I - \eta_t H^{[t]}$, whose products govern how a data point's influence evolves through subsequent optimization steps. That analysis was agnostic to whether influence grows or decays; the WAIM estimator is valid in either regime. We now ask: under what conditions does influence decay naturally, and can this decay be exploited for certified unlearning?

The setting we analyze is single-pass mini-batch SGD with coupled L_2 regularization (i.e., the regularization term $\frac{\lambda}{2}\|\boldsymbol{\theta}\|^2$ is added to the loss before differentiation). Coupled L_2 regularization shifts the Hessian spectrum by λI, but does not by itself guarantee positive definiteness in non-convex models. Under Assumption 4 and the step-size condition $\eta_t \leq 1/L$, where L is the Hessian upper bound, the propagator satisfies $\|Z_t\|_2 \leq 1 - \eta_t m < 1$ at every step, producing geometric contraction. The influence of any single data point therefore shrinks exponentially with each subsequent training step. This natural decay transforms the unlearning problem: rather than explicitly computing and subtracting a data point's accumulated influence (which requires stored gradients, Hessian information, and trajectory replay), one can inject calibrated Gaussian noise to mask whatever residual influence remains [7, 10, 12]. The noise magnitude decays exponentially with the propagation interval, making the utility impact negligible for data points that have been in the training pipeline long enough.

The scope of the analysis is subject to two important limitations. First, coupled L_2 regularization differs from the decoupled weight decay used in AdamW [13], the optimizer of choice for most large language model training. The contraction analysis relies on the λI shift appearing in the Hessian, which holds for coupled regularization but requires separate treatment for decoupled implementations. Second, the theoretical guarantees are derived without experimental validation; the framework is a mathematical prototype whose practical viability in large-scale settings remains to be established. We state these limitations explicitly so that the theoretical contribution, the principle that natural influence decay under regularization can substitute for explicit influence computation in certified unlearning, is evaluated on its own terms.

The contributions of this chapter are threefold. First, we formalize exponential influence decay under a local contractivity condition for single-pass SGD with coupled L_2 regularization, extending the propagator analysis of Chap. 4 from a descriptive tool to a prescriptive one. Second, we design a decay-aware certified unlearning mechanism that replaces explicit influence computation with calibrated Gaussian noise, achieving (ε, δ)-certified unlearning relative to a noisy retraining oracle with $O(p)$ per-request cost. Third, we provide theoretical guarantees characterizing the required noise magnitude, the request-specific maturity condition needed for utility preservation, and the complexity advantages over baseline approaches.

System Model and Problem Statement

This section establishes the training dynamics that govern how data influence evolves (Section "System Model") and formalizes the online certified unlearning problem that the remainder of this chapter addresses (Section "Problem Statement").

System Model

Training Framework

The training setup follows Chap. 4 in a simplified SGD-like setting. The training set is $\mathcal{D} = \{z_i\}_{i=1}^N$ with $z_i = (x_i, y_i) \in \mathcal{Z} = \mathcal{X} \times \mathcal{Y}$, and the model $f : \mathcal{X} \times \Theta \to \mathcal{Y}$ is parameterized by $\boldsymbol{\theta} \in \Theta \subseteq \mathbb{R}^p$. Training proceeds by mini-batch SGD over T steps: at step t, a mini-batch $S_t \subseteq \{1, \ldots, N\}$ of size $B = |S_t|$ is sampled, and parameters are updated as

$$\boldsymbol{\theta}^{[t+1]} = \boldsymbol{\theta}^{[t]} - \frac{\eta_t}{|S_t|} \sum_{i \in S_t} g(z_i; \boldsymbol{\theta}^{[t]}), \tag{6.1}$$

where $g(z; \boldsymbol{\theta}) = \nabla_{\boldsymbol{\theta}} \ell(z; \boldsymbol{\theta})$ is the per-sample gradient and η_t is the learning rate.

The per-sample loss incorporates *coupled* L_2 regularization of strength $\lambda > 0$ [13, 14], so

$$\ell(z; \boldsymbol{\theta}) = \ell_0(z; \boldsymbol{\theta}) + \frac{\lambda}{2} \|\boldsymbol{\theta}\|^2,$$

where ℓ_0 is the original loss. Throughout this chapter, we analyze this coupled-regularization SGD-like model directly; decoupled weight decay and adaptive optimizers such as AdamW are outside the scope of the formal analysis.

For a fixed mini-batch S_t, define the mini-batch Hessian at parameter value $\boldsymbol{\vartheta}$ by

$$H^{[t]}(\boldsymbol{\vartheta}) = \frac{1}{|S_t|} \sum_{i \in S_t} \nabla^2_{\boldsymbol{\theta}} \ell(z_i; \boldsymbol{\vartheta}).$$

The linearized propagator matrix at step t is $Z_t = I - \eta_t H^{[t]}(\boldsymbol{\theta}^{[t]})$.

Under single-pass training, each data point z_j appears in exactly one mini-batch at step t_{z_j}. We couple the original run and the deleted run by using the same mini-batch sequence and optimizer randomness; at step t_{z_j}, the deleted run omits the direct contribution of z_j while keeping the mini-batch normalization fixed at $1/B$. With this coupling, the initial parameter deviation is

$$\Delta\boldsymbol{\theta}_{-j}^{[t_{z_j}+1]} = \boldsymbol{\theta}_{-j}^{[t_{z_j}+1]} - \boldsymbol{\theta}^{[t_{z_j}+1]} = \frac{\eta_{t_{z_j}}}{B}\, g(z_j;\, \boldsymbol{\theta}^{[t_{z_j}]}), \tag{6.2}$$

and for later steps the deviation propagates approximately as

$$\Delta\boldsymbol{\theta}_{-j}^{[t+1]} \approx Z_t\, \Delta\boldsymbol{\theta}_{-j}^{[t]}, \qquad t > t_{z_j},$$

as in Chap. 4.

Assumptions

The analysis requires three structural conditions on the training dynamics.

Assumption 4 (*Local Spectral Bounds/Local Contractivity*) There exist constants $0 < m \leq L < \infty$ such that for every training step t and every point $\boldsymbol{\vartheta}$ on the line segment joining $\boldsymbol{\theta}^{[t]}$ and $\boldsymbol{\theta}_{-j}^{[t]}$,

$$m\, I \preceq H^{[t]}(\boldsymbol{\vartheta}) \preceq L\, I.$$

This is a *local* contractivity assumption along the deletion path. In nonconvex models, coupled L_2 regularization shifts the Hessian spectrum by λI and can improve the lower bound, but it does not by itself guarantee $m > 0$ globally. We therefore use Assumption 4 as the formal condition under which the decay analysis is valid.

Assumption 5 (*Learning Rate Bound*) The learning rate satisfies $\eta_t \leq \frac{1}{L}$ for all t.

This ensures that the exact one-step Jacobian is non-expansive in every eigendirection and contractive by at least $1 - \eta_t m$ under Assumption 4.

Assumption 6 (*Bounded Gradients*) There exists $G > 0$ such that $\|g(z;\, \boldsymbol{\theta})\| \leq G$ for all $z \in \mathcal{D}$ and all $\boldsymbol{\theta}$ on the training trajectory.

This bounds the initial influence of any data point:

$$\|\Delta\boldsymbol{\theta}_{-j}^{[t_{z_j}+1]}\| \leq \frac{\eta_{t_{z_j}}}{B}\, G.$$

Per-sample clipping enforces such a bound exactly; in large-model practice it is best viewed as an analytical assumption or an approximation induced by clipping and bounded-activation regimes.

Problem Statement

The aim of this chapter is to design an unlearning operator $\mathcal{U}$ that can remove the influence of a training data point once its residual influence has decayed sufficiently, without halting or restarting the training process, while preserving model quality and requiring negligible computational overhead.

Unlike the post-training setting studied in earlier chapters, the online setting introduces a temporal dimension. A data point z_j enters the training process at step t_{z_j}, the unique step at which the mini-batch containing z_j is used. An unlearning request arrives at some later step $\tau \geq t_{z_j} + 1$, demanding removal of z_j's influence from the current model state $\boldsymbol{\theta}^{[\tau]}$. The propagation interval

$$\Delta = \tau - t_{z_j} - 1$$

measures how many optimization steps have elapsed since z_j was used.

Because the unlearning mechanism will inject Gaussian noise, the natural comparison object is a *noisy retraining oracle* rather than exact retraining. Let

$$\tilde{\boldsymbol{\theta}}^{[\tau]}_{\text{retrain}} = \boldsymbol{\theta}^{[\tau]}_{-j} + \boldsymbol{\xi}', \qquad \boldsymbol{\xi}' \sim \mathcal{N}(\mathbf{0}, \sigma^2 I), \tag{6.3}$$

where $\boldsymbol{\theta}^{[\tau]}_{-j}$ is the parameter vector obtained by retraining on $\mathcal{D} \setminus \{z_j\}$ under the same coupling.

The operator $\mathcal{U}$ must satisfy four requirements simultaneously, formalized as follows.

Definition 6.1 (*Online Certified Unlearning Problem*) Given the SGD process (Eq. (6.1)) with coupled L_2 regularization, an unlearning request for data point z_j used at step t_{z_j}, and a privacy budget (ε, δ), construct an operator $\mathcal{U}$ satisfying:

1. **Certified removal**. The unlearned parameters $\tilde{\boldsymbol{\theta}}^{[\tau]} = \mathcal{U}(\boldsymbol{\theta}^{[\tau]}, z_j)$ are (ε, δ)-indistinguishable from the noisy retraining oracle in Eq. (6.3).
2. **Utility preservation**. The expected excess empirical risk relative to the *same-noise* retraining baseline is bounded:

$$\mathbb{E}\big[\hat{R}(\tilde{\boldsymbol{\theta}}^{[\tau]}; \mathcal{D} \setminus \{z_j\})\big] - \mathbb{E}\big[\hat{R}(\tilde{\boldsymbol{\theta}}^{[\tau]}_{\text{retrain}}; \mathcal{D} \setminus \{z_j\})\big] \leq \xi, \tag{6.4}$$

   for a user-specified tolerance $\xi > 0$.
3. **Computational efficiency**. The time and space complexity of $\mathcal{U}$ is $o(T \cdot B \cdot p)$, strictly sublinear in the cost of retraining.
4. **Online capability**. The operator $\mathcal{U}$ requires only the current parameters $\boldsymbol{\theta}^{[\tau]}$ and request metadata sufficient to recover t_{z_j} (equivalently, Δ), without stored gradients, Hessians, or the full training trajectory.

These four requirements create a natural tension. Certified removal demands that the residual influence of z_j be masked, either by explicit subtraction or by noise injection. Noise injection can degrade utility, while explicit subtraction incurs a computational cost proportional to the training trajectory. The key insight of this chapter is that local contractivity resolves this tension: the influence of z_j at step τ decays exponentially with the propagation interval Δ, so the noise required for certification becomes exponentially smaller over time. Once Δ exceeds a request-specific maturity threshold $\Delta_{\min,j}$ (derived in Section "Maturity Condition"), a single noise injection satisfies all four requirements simultaneously without any explicit influence computation.

In-Training Unlearning Framework

Under local contractivity, the influence of each data point decays exponentially through subsequent optimization steps. After enough steps, the residual influence is small enough to be masked by a small amount of Gaussian noise, yielding certified unlearning without any explicit influence computation. This section presents the resulting unlearning mechanism: Section "Influence Decay and Noise Injection" describes the influence decay property and the noise injection operator, Section "Maturity Condition" establishes the request-specific maturity condition that ensures both certification and utility preservation, and Section "Complete Unlearning Protocol" assembles the complete protocol. All formal proofs appear in Section "Theoretical Analysis".

Influence Decay and Noise Injection

Under Assumptions 4 and 5, the exact one-step propagator along the deletion path is uniformly contractive. Section "Influence Decay Analysis" shows that for an appropriate path-averaged Hessian $\bar{H}^{[t]}$,

$$\|I - \eta_t \bar{H}^{[t]}\|_2 \;\leq\; 1 - \eta_t m \;\leq\; 1 - \eta_{\min} m \;=:\; \rho \;<\; 1, \tag{6.5}$$

where $\eta_{\min} = \min_t \eta_t$. Coupled L_2 regularization helps by shifting the Hessian spectrum, but the decay result ultimately relies on the explicit local contractivity condition in Assumption 4.

This contraction causes the influence of any single data point to decay exponentially. A data point z_j used at step t_{z_j} produces an initial deviation of magnitude at most

$$c_j = \frac{\eta_{t_{z_j}}}{B} G.$$

Since each subsequent propagator contracts by at most ρ, the residual influence at step τ satisfies

$$\|\Delta\boldsymbol{\theta}_{-j}^{[\tau]}\| \;\leq\; \rho^{\Delta} \cdot c_j, \tag{6.6}$$

where $\Delta = \tau - t_{z_j} - 1$ is the propagation interval.

To convert this natural decay into certified unlearning, the framework injects isotropic Gaussian noise calibrated to the residual influence. Upon receiving an unlearning request for z_j at step τ, the operator produces

$$\tilde{\boldsymbol{\theta}}^{[\tau]} = \mathcal{U}(\boldsymbol{\theta}^{[\tau]}, z_j) = \boldsymbol{\theta}^{[\tau]} + \boldsymbol{\xi}, \qquad \boldsymbol{\xi} \sim \mathcal{N}(\mathbf{0}, \sigma_j^2 I), \tag{6.7}$$

with noise standard deviation

$$\sigma_j = \rho^{\Delta} \cdot c_j \cdot \frac{\sqrt{2\ln(1.25/\delta)}}{\varepsilon}. \tag{6.8}$$

The noise magnitude is the product of the residual influence bound $\rho^{\Delta} c_j$ and a privacy calibration constant depending on (ε, δ). This calibration ensures (ε, δ)-certified unlearning relative to the noisy retraining oracle (proved in Section "Certified Unlearning"). The operator requires no gradient computation, no Hessian inversion, and no trajectory replay; its cost is $O(p)$.

Maturity Condition

The noise standard deviation in Eq. (6.8) decays exponentially with the propagation interval Δ, but for small Δ the residual influence may still be too large to preserve utility within a given tolerance. Section "Utility Preservation" shows that the excess empirical risk of the unlearned model relative to the noisy retraining baseline is bounded by

$$G \cdot \rho^{\Delta} \cdot c_j + \frac{L}{2}\,\rho^{2\Delta}\,c_j^2.$$

To ensure that this quantity is at most a user-specified tolerance $\xi > 0$, let

$$y^{\star}(\xi; G, L) = \begin{cases} \xi/G, & L = 0, \\ \dfrac{\sqrt{G^2 + 2L\xi} - G}{L}, & L > 0. \end{cases} \tag{6.9}$$

Then the sufficient and necessary condition for the quadratic bound to be below ξ is

$$\rho^{\Delta} c_j \leq y^{\star}.$$

Equivalently, the request-specific maturity threshold is

$$\Delta_{\min,j} = \begin{cases} 0, & c_j \le y^\star, \\ \left\lceil \dfrac{\ln(c_j/y^\star)}{\ln(1/\rho)} \right\rceil, & c_j > y^\star. \end{cases} \tag{6.10}$$

For small utility tolerances ξ, one has $y^\star = \xi/G + O(\xi^2)$, so Eq. (6.10) is well approximated by

$$\Delta_{\min,j} \approx \left\lceil \frac{\ln(Gc_j/\xi)}{\ln(1/\rho)} \right\rceil. \tag{6.11}$$

This approximation recovers the simpler expression used for back-of-the-envelope estimates.

The maturity condition $\Delta \ge \Delta_{\min,j}$ is the sole precondition for applying the unlearning operator to request j. An unlearning request for data point z_j is therefore eligible for execution at any step

$$\tau \ge t_{z_j} + 1 + \Delta_{\min,j}.$$

Before this point, the influence of z_j has not yet decayed enough, and the system simply waits for training to progress naturally. No explicit influence computation, gradient storage, or trajectory replay is needed at any stage.

Table 6.1 illustrates the approximate threshold in Eq. (6.11) for representative contractive training configurations.

For representative contractive parameters ($m = 0.01$, $\eta = 3 \times 10^{-4}$, $B = 1024$, $G = 1.0$, $\xi = 10^{-7}$), the approximate threshold is approximately 3.5×10^5 steps. In a multi-million-step training run, this represents a moderate fraction of the total duration. Whether this is acceptable depends on the timing of the deletion request: requests for older samples mature naturally, while very recent samples may remain ineligible until sufficient contraction has accumulated.

Table 6.1 Approximate minimum propagation threshold $\Delta_{\min}$ for representative contractive configurations, using Eq. (6.11). Fixed parameters: $B = 1024$, $G = 1.0$. Here $c_j = \eta G/B$ and $\rho = 1 - \eta m$

Curvature floor m	Learning rate η	$c_j = \eta G/B$	ρ	$\xi = 10^{-7}$	$\xi = 10^{-8}$
0.01	3×10^{-4}	2.93×10^{-7}	0.999997	3.6×10^5	1.1×10^6
0.01	1×10^{-3}	9.77×10^{-7}	0.99999	2.3×10^5	4.6×10^5
0.1	3×10^{-4}	2.93×10^{-7}	0.99997	3.6×10^4	1.1×10^5
0.1	1×10^{-3}	9.77×10^{-7}	0.9999	2.3×10^4	4.6×10^4

Complete Unlearning Protocol

Algorithm 8 formalizes the noise injection procedure for a single unlearning request that has satisfied its request-specific maturity condition.

Algorithm 8 Decay-Aware Noise Injection for One Mature Request

Require: Current step τ, parameters $\boldsymbol{\theta}^{[\tau]}$, target z_j used at step t_{z_j}

Require: Contraction rate ρ, initial deviation bound $c_j = \frac{\eta_{t_{z_j}}}{B} G$, privacy budget (ε, δ)

Require: Maturity condition satisfied: $\tau - t_{z_j} - 1 \geq \Delta_{\min,j}$

Ensure: Unlearned parameters $\tilde{\boldsymbol{\theta}}$

1: Compute propagation interval: $\Delta \leftarrow \tau - t_{z_j} - 1$

2: Compute noise scale:

$$\sigma_j \leftarrow \rho^{\Delta} \cdot c_j \cdot \sqrt{2\ln(1.25/\delta)} \,/\, \varepsilon$$

3: Sample $\boldsymbol{\xi} \sim \mathcal{N}(\mathbf{0}, \sigma_j^2 I)$

4: **return** $\tilde{\boldsymbol{\theta}} \leftarrow \boldsymbol{\theta}^{[\tau]} + \boldsymbol{\xi}$

The procedure is deliberately minimal: compute the propagation interval, calibrate the noise, and perturb the parameters. No gradient computation, Hessian inversion, or trajectory replay is involved. The entire operation costs $O(p)$.

When k mature unlearning requests are processed at the same step, each receives independently calibrated noise:

$$\tilde{\boldsymbol{\theta}}^{[\tau]} = \boldsymbol{\theta}^{[\tau]} + \sum_{i=1}^{k} \boldsymbol{\xi}_i, \qquad \boldsymbol{\xi}_i \sim \mathcal{N}\left(\mathbf{0}, \sigma_i^2 I\right), \tag{6.12}$$

where

$$\sigma_i = \rho^{\Delta_i} \cdot c_i \cdot \frac{\sqrt{2\ln(1.25/\delta_i)}}{\varepsilon_i}.$$

The joint privacy guarantee follows from composition [15]: under basic composition, $\varepsilon_{\text{total}} = \sum_i \varepsilon_i$ and $\delta_{\text{total}} = \sum_i \delta_i$, with tighter accounting available when many requests are processed. Since each σ_i^2 decays exponentially with Δ_i, the cumulative utility impact is dominated by the few most recent requests.

Algorithm 9 shows how the unlearning mechanism integrates into the training loop. Each request stores the sample-use step t_{z_j}, its initial deviation bound c_j, and its maturity time

$$m_j = t_{z_j} + 1 + \Delta_{\min,j}.$$

At each training step, the loop processes every request whose maturity time has been reached; training then proceeds as usual. Because all subsequent updates are post-processing of the noised state under the same coupled randomness, the certification guarantee continues to hold at later checkpoints.

Algorithm 9 Training Loop with Decay-Aware Unlearning

Require: Dataset $\mathcal{D}$, total steps T, utility tolerance ξ
Initialize parameters $\boldsymbol{\theta}^{[0]}$, pending requests $Q \leftarrow \emptyset$
for $t = 0$ **to** $T - 1$ **do**
  Receive any new unlearning requests that arrive at step t
  for each new request targeting z_j **do**
    Recover or look up the step index t_{z_j}
    Compute $c_j \leftarrow \frac{\eta_{t_{z_j}}}{B} G$
    Compute $\Delta_{\min,j}$ from Eq. (6.10)
    Set maturity time $m_j \leftarrow t_{z_j} + 1 + \Delta_{\min,j}$
    Insert $(z_j, t_{z_j}, c_j, m_j, \varepsilon_j, \delta_j)$ into Q
  end for
  {**Phase 1: Process mature unlearning requests**}
  for each $(z_j, t_{z_j}, c_j, m_j, \varepsilon_j, \delta_j) \in Q$ with $m_j \leq t$ **do**
    $\boldsymbol{\theta}^{[t]} \leftarrow \text{DecayAwareNoiseInjection}(\boldsymbol{\theta}^{[t]}, t, z_j, t_{z_j}, c_j, \rho, \varepsilon_j, \delta_j)$
    Remove $(z_j, t_{z_j}, c_j, m_j, \varepsilon_j, \delta_j)$ from Q
  end for
  {**Phase 2: Standard SGD update**}
  Sample mini-batch S_t
  $\boldsymbol{\theta}^{[t+1]} \leftarrow \boldsymbol{\theta}^{[t]} - \frac{\eta_t}{|S_t|} \sum_{i \in S_t} g(z_i; \boldsymbol{\theta}^{[t]})$
end for

The overhead of the unlearning integration is negligible relative to training. Request processing requires storing or recovering t_{z_j}, computing the request-specific maturity time, and adding a Gaussian vector when the request matures. If pending requests are organized by maturity time, the training-loop overhead is small compared with the $O(B \cdot p)$ cost of a standard SGD step.

Theoretical Analysis

The preceding section presented the decay-aware unlearning framework and stated the key bounds that govern its operation: the exponential decay of data influence (Eq. (6.6)), the noise calibration for certified removal (Eq. (6.8)), and the request-specific maturity condition for utility preservation (Eq. (6.10)). This section provides the formal proofs underlying these claims and analyzes their implications. Section "Influence Decay Analysis" establishes the influence decay bound, first under the linearized propagation model and then under the exact nonlinear dynamics. Section "Certified Unlearning" proves that the noise injection mechanism achieves (ε, δ)-certified unlearning relative to the noisy retraining oracle. Section "Utility Preservation" derives the utility preservation guarantee and the request-specific propagation threshold. Section "Complexity Analysis" provides a formal complexity analysis comparing the framework against baseline methods.

Influence Decay Analysis

The exponential decay of data influence is the theoretical cornerstone of the framework. We establish this result in two stages: a clean bound under the linearized influence propagation model, followed by an exact bound under the local contractivity assumption.

Linearized Influence Decay

Recall from Chap. 4 that the parameter deviation

$$\Delta\boldsymbol{\theta}_{-j}^{[t]} = \boldsymbol{\theta}_{-j}^{[t]} - \boldsymbol{\theta}^{[t]}$$

measures the cumulative effect of excluding data point z_j from the training process. Under single-pass training, z_j appears in exactly one mini-batch at step t_{z_j}, producing an initial deviation

$$\Delta\boldsymbol{\theta}_{-j}^{[t_{z_j}+1]} = \frac{\eta_{t_{z_j}}}{B}\, g(z_j;\, \boldsymbol{\theta}^{[t_{z_j}]}). \tag{6.13}$$

For all subsequent steps $t > t_{z_j}$, the linearized influence propagation recursion (Chap. 4, Eq. (4.14)) reduces to

$$\Delta\boldsymbol{\theta}_{-j}^{[t+1]} \approx Z_t\, \Delta\boldsymbol{\theta}_{-j}^{[t]}, \qquad t > t_{z_j}, \tag{6.14}$$

since $z_j \notin S_t$ for $t \neq t_{z_j}$ and the direct contribution term vanishes. Unrolling from step $t_{z_j} + 1$ to step τ yields

$$\Delta\boldsymbol{\theta}_{-j}^{[\tau]} \approx \left(\prod_{s=t_{z_j}+1}^{\tau-1} Z_s \right) \Delta\boldsymbol{\theta}_{-j}^{[t_{z_j}+1]}. \tag{6.15}$$

Proposition 6.1 (Linearized Influence Decay) *Under Assumptions 4 and 5, the linearized parameter deviation satisfies*

$$\left\| \prod_{s=t_{z_j}+1}^{\tau-1} Z_s\, \Delta\boldsymbol{\theta}_{-j}^{[t_{z_j}+1]} \right\| \leq \rho^{\Delta} \cdot \frac{\eta_{t_{z_j}}}{B}\, G, \tag{6.16}$$

where $\Delta = \tau - t_{z_j} - 1$ *and* $\rho = 1 - \eta_{min}\, m$.

Proof For any s, Assumptions 4 and 5 imply that all eigenvalues of $H^{[s]}(\boldsymbol{\theta}^{[s]})$ lie in $[m, L]$ and $\eta_s \le 1/L$. Hence all eigenvalues of

$$Z_s = I - \eta_s H^{[s]}(\boldsymbol{\theta}^{[s]})$$

lie in $[0, 1 - \eta_s m]$, so

$$\|Z_s\|_2 \le 1 - \eta_s m \le 1 - \eta_{\min} m = \rho.$$

By submultiplicativity of the spectral norm,

$$\left\| \prod_{s=t_{z_j}+1}^{\tau-1} Z_s \right\|_2 \le \prod_{s=t_{z_j}+1}^{\tau-1} \|Z_s\|_2 \le \rho^{\Delta}.$$

The initial deviation satisfies

$$\|\Delta\boldsymbol{\theta}_{-j}^{[t_{z_j}+1]}\| = \frac{\eta_{t_{z_j}}}{B}\|g(z_j;\, \boldsymbol{\theta}^{[t_{z_j}]})\| \le \frac{\eta_{t_{z_j}}}{B} G$$

by Assumption 6. Combining the two bounds yields the result. □

Exact Influence Decay Under Local Contractivity

Proposition 6.1 uses the point Hessian $H^{[t]}(\boldsymbol{\theta}^{[t]})$. We now show that under Assumption 4, the same exponential decay bound holds for the exact nonlinear dynamics.

For $t > t_{z_j}$ (so that $z_j \notin S_t$), the exact deviation evolves as

$$\Delta\boldsymbol{\theta}_{-j}^{[t+1]} = \Delta\boldsymbol{\theta}_{-j}^{[t]} - \frac{\eta_t}{B}\sum_{i\in S_t}\big[g(z_i;\, \boldsymbol{\theta}_{-j}^{[t]}) - g(z_i;\, \boldsymbol{\theta}^{[t]})\big]. \tag{6.17}$$

Applying the integral form of the mean value theorem to each gradient difference yields

$$g(z_i;\, \boldsymbol{\theta}_{-j}^{[t]}) - g(z_i;\, \boldsymbol{\theta}^{[t]}) = \bar{H}_i^{[t]}\, \Delta\boldsymbol{\theta}_{-j}^{[t]}, \tag{6.18}$$

where

$$\bar{H}_i^{[t]} = \int_0^1 \nabla_\theta^2\, \ell\Big(z_i;\, \boldsymbol{\theta}^{[t]} + s\,\Delta\boldsymbol{\theta}_{-j}^{[t]}\Big)\, ds.$$

Define the path-averaged mini-batch Hessian

$$\bar{H}^{[t]} = \frac{1}{B}\sum_{i\in S_t} \bar{H}_i^{[t]}.$$

Then the exact recursion becomes

$$\Delta\boldsymbol{\theta}_{-j}^{[t+1]} = (I - \eta_t\,\bar{H}^{[t]})\,\Delta\boldsymbol{\theta}_{-j}^{[t]}. \tag{6.19}$$

Theorem 6.1 (Exact Influence Decay) *Under Assumptions 4–6, define*

$$c_j = \frac{\eta_{t_{z_j}}}{B} G \quad \text{and} \quad \rho = 1 - \eta_{min} m.$$

Then the exact parameter deviation satisfies

$$\|\Delta\boldsymbol{\theta}_{-j}^{[\tau]}\| \;\le\; \rho^{\Delta} \cdot c_j \tag{6.20}$$

for all $\tau \ge t_{z_j} + 1$, *where* $\Delta = \tau - t_{z_j} - 1$.

Proof By Assumption 4, every Hessian on the segment joining $\boldsymbol{\theta}^{[t]}$ and $\boldsymbol{\theta}_{-j}^{[t]}$ has eigenvalues in $[m, L]$. Since $\bar{H}^{[t]}$ is an average of such Hessians, it also satisfies

$$mI \preceq \bar{H}^{[t]} \preceq LI.$$

Because $\eta_t \le 1/L$, every eigenvalue of $I - \eta_t \bar{H}^{[t]}$ lies in $[0, 1 - \eta_t m]$, so

$$\|I - \eta_t \bar{H}^{[t]}\|_2 \le 1 - \eta_t m \le 1 - \eta_{\min} m = \rho.$$

Taking norms in Eq. (6.19) gives

$$\|\Delta\boldsymbol{\theta}_{-j}^{[t+1]}\| \le \rho\, \|\Delta\boldsymbol{\theta}_{-j}^{[t]}\|.$$

Iterating this recursion from step $t_{z_j} + 1$ to step τ yields

$$\|\Delta\boldsymbol{\theta}_{-j}^{[\tau]}\| \le \rho^{\Delta} \|\Delta\boldsymbol{\theta}_{-j}^{[t_{z_j}+1]}\|.$$

Finally, Assumption 6 and Eq. (6.13) imply

$$\|\Delta\boldsymbol{\theta}_{-j}^{[t_{z_j}+1]}\| \le \frac{\eta_{t_{z_j}}}{B} G = c_j,$$

which proves the claim. □

The theorem shows that the exact nonlinear dynamics obey the same exponential decay pattern as the linearized model, provided the optimization remains locally contractive along the deletion path. For representative values $B = 1024$, $\eta = 3 \times 10^{-4}$, and $G = 1.0$, the initial deviation scale is $c_j \approx 2.9 \times 10^{-7}$. If the local curvature floor is $m = 0.01$, then $\rho \approx 0.999997$, implying slow but steady geometric decay.

Certified Unlearning

We now prove that the noise injection operator

$$\mathcal{U}(\boldsymbol{\theta}^{[\tau]}, z_j) = \boldsymbol{\theta}^{[\tau]} + \boldsymbol{\xi}, \qquad \boldsymbol{\xi} \sim \mathcal{N}(\mathbf{0}, \sigma_j^2 I)$$

with σ_j calibrated according to Eq. (6.8) achieves (ε, δ)-certified unlearning relative to the noisy retraining oracle.

Noisy Retraining Baseline

The ideal retrained model is $\boldsymbol{\theta}_{-j}^{[\tau]}$, the parameters that would result from running the coupled training procedure on $\mathcal{D} \setminus \{z_j\}$ under the same mini-batch and randomness coupling. The noisy retrained model is

$$\mathcal{R}_{\sigma_j}(\mathcal{D} \setminus \{z_j\}) = \boldsymbol{\theta}_{-j}^{[\tau]} + \boldsymbol{\xi}', \qquad \boldsymbol{\xi}' \sim \mathcal{N}(\mathbf{0}, \sigma_j^2 I), \tag{6.21}$$

with the same noise variance σ_j^2. Both

$$\mathcal{U}(\boldsymbol{\theta}^{[\tau]}, z_j) \quad \text{and} \quad \mathcal{R}_{\sigma_j}(\mathcal{D} \setminus \{z_j\})$$

are Gaussian random vectors with identical covariance $\sigma_j^2 I$, differing only in their means by

$$\boldsymbol{d} = \boldsymbol{\theta}^{[\tau]} - \boldsymbol{\theta}_{-j}^{[\tau]} = -\Delta\boldsymbol{\theta}_{-j}^{[\tau]}.$$

Certification Guarantee

Theorem 6.2 (Certified Unlearning via Noise Injection) *Under the conditions of Theorem 6.1, let* $\varepsilon \in (0, 1)$ *and* $\delta \in (0, 1)$. *Define the noise standard deviation*

$$\sigma_j = \rho^{\Delta} \cdot c_j \cdot \frac{\sqrt{2\ln(1.25/\delta)}}{\varepsilon}. \tag{6.22}$$

Then the unlearning operator $\mathcal{U}(\boldsymbol{\theta}^{[\tau]}, z_j) = \boldsymbol{\theta}^{[\tau]} + \mathcal{N}(\mathbf{0}, \sigma_j^2 I)$ *achieves* (ε, δ)-*certified unlearning relative to the noisy retraining oracle* $\mathcal{R}_{\sigma_j}(\mathcal{D} \setminus \{z_j\})$.

Proof Condition on any realization of the mini-batch sequence and the optimizer randomness used to couple the original and deleted runs. Under this conditioning,

$$\boldsymbol{d} = \boldsymbol{\theta}^{[\tau]} - \boldsymbol{\theta}_{-j}^{[\tau]}$$

is deterministic and satisfies

$$\|\boldsymbol{d}\| = \|\Delta\boldsymbol{\theta}_{-j}^{[\tau]}\| \le \rho^{\Delta} c_j$$

by Theorem 6.1. This quantity is the ℓ_2-sensitivity between the two coupled training outputs.

Conditioned on the same randomness, the unlearned model is distributed as

$$\mathcal{N}(\boldsymbol{\theta}^{[\tau]}, \sigma_j^2 I)$$

and the noisy retrained model as

$$\mathcal{N}(\boldsymbol{\theta}_{-j}^{[\tau]}, \sigma_j^2 I).$$

By the standard Gaussian mechanism analysis [15], if

$$\sigma_j \geq \frac{\|\boldsymbol{d}\|\sqrt{2\ln(1.25/\delta)}}{\varepsilon},$$

then these two Gaussian distributions are (ε, δ)-indistinguishable. Since

$$\|\boldsymbol{d}\| \leq \rho^{\Delta} c_j,$$

the choice in Eq. (6.22) suffices.

Finally, integrating the conditional indistinguishability inequality over the randomness of the mini-batch sequence and optimizer coupling yields the same (ε, δ) guarantee unconditionally. The reverse direction follows symmetrically, completing the two-sided guarantee. □

The exponential dependence of σ_j on Δ is the central feature of this result. As the propagation interval grows, the noise required for certification decays at the same rate as the residual influence. Each parameter coordinate receives Gaussian noise with standard deviation σ_j, while the ℓ_2 magnitude of the full noise vector scales on the order of $\sigma_j\sqrt{p}$. Utility should therefore be assessed through the comparison to the same-noise retraining baseline, rather than by interpreting the perturbation coordinatewise in isolation.

Utility Preservation

Certification ensures statistical indistinguishability from the noisy retraining oracle, but practitioners also require that the unlearned model perform well on the retained data $\mathcal{D}' = \mathcal{D} \setminus \{z_j\}$. This subsection bounds the excess risk of the unlearned model relative to the *same-noise* retraining baseline.

Excess Risk Bound

Theorem 6.3 (Utility Preservation Relative to Noisy Retraining) *Under the conditions of Theorem 6.1, let*

$$\tilde{\boldsymbol{\theta}}^{[\tau]} = \boldsymbol{\theta}^{[\tau]} + \boldsymbol{\xi} \quad \textit{and} \quad \tilde{\boldsymbol{\theta}}^{[\tau]}_{-j} = \boldsymbol{\theta}^{[\tau]}_{-j} + \boldsymbol{\xi}',$$

where $\boldsymbol{\xi}, \boldsymbol{\xi}' \sim \mathcal{N}(\mathbf{0}, \sigma_j^2 I)$ *independently. Suppose the empirical risk*

$$\hat{R}(\boldsymbol{\theta};\, \mathcal{D}') = \frac{1}{|\mathcal{D}'|} \sum_{z \in \mathcal{D}'} \ell(z;\, \boldsymbol{\theta})$$

has L*-Lipschitz continuous gradients. Then*

$$\mathbb{E}\big[\hat{R}(\tilde{\boldsymbol{\theta}}^{[\tau]};\, \mathcal{D}')\big] - \mathbb{E}\big[\hat{R}(\tilde{\boldsymbol{\theta}}^{[\tau]}_{-j};\, \mathcal{D}')\big] \;\le\; G\,\rho^{\Delta}\,c_j + \frac{L}{2}\,\rho^{2\Delta}\,c_j^2. \tag{6.23}$$

Proof Define the smoothed empirical risk

$$F_{\sigma_j}(\boldsymbol{\theta};\, \mathcal{D}') = \mathbb{E}_{\boldsymbol{\zeta} \sim \mathcal{N}(\mathbf{0}, \sigma_j^2 I)}\big[\hat{R}(\boldsymbol{\theta} + \boldsymbol{\zeta};\, \mathcal{D}')\big].$$

Then

$$\mathbb{E}\big[\hat{R}(\tilde{\boldsymbol{\theta}}^{[\tau]};\, \mathcal{D}')\big] = F_{\sigma_j}(\boldsymbol{\theta}^{[\tau]};\, \mathcal{D}')$$

and

$$\mathbb{E}\big[\hat{R}(\tilde{\boldsymbol{\theta}}^{[\tau]}_{-j};\, \mathcal{D}')\big] = F_{\sigma_j}(\boldsymbol{\theta}^{[\tau]}_{-j};\, \mathcal{D}').$$

Since expectation preserves smoothness, $F_{\sigma_j}(\cdot;\, \mathcal{D}')$ is also L-smooth. Moreover,

$$\nabla F_{\sigma_j}(\boldsymbol{\theta};\, \mathcal{D}') = \mathbb{E}_{\boldsymbol{\zeta}}\big[\nabla \hat{R}(\boldsymbol{\theta} + \boldsymbol{\zeta};\, \mathcal{D}')\big].$$

By Assumption 6, each per-sample gradient has norm at most G, hence the average gradient and its Gaussian smoothing also satisfy

$$\|\nabla F_{\sigma_j}(\boldsymbol{\theta};\, \mathcal{D}')\| \le G.$$

Let

$$\boldsymbol{d} = \boldsymbol{\theta}^{[\tau]} - \boldsymbol{\theta}^{[\tau]}_{-j}.$$

Then by L-smoothness of F_{σ_j},

$$F_{\sigma_j}(\boldsymbol{\theta}^{[\tau]}_{-j} + \boldsymbol{d};\, \mathcal{D}') \le F_{\sigma_j}(\boldsymbol{\theta}^{[\tau]}_{-j};\, \mathcal{D}') + \nabla F_{\sigma_j}(\boldsymbol{\theta}^{[\tau]}_{-j};\, \mathcal{D}')^\top \boldsymbol{d} + \frac{L}{2}\|\boldsymbol{d}\|^2.$$

Using Cauchy–Schwarz and $\|\nabla F_{\sigma_j}\| \leq G$ gives

$$F_{\sigma_j}(\boldsymbol{\theta}^{[\tau]}; \mathcal{D}') - F_{\sigma_j}(\boldsymbol{\theta}_{-j}^{[\tau]}; \mathcal{D}') \leq G\|\boldsymbol{d}\| + \frac{L}{2}\|\boldsymbol{d}\|^2.$$

Finally, Theorem 6.1 implies

$$\|\boldsymbol{d}\| = \|\Delta\boldsymbol{\theta}_{-j}^{[\tau]}\| \leq \rho^{\Delta} c_j,$$

which yields Eq. (6.23). □

The bound consists of two terms, both decaying exponentially with Δ. The first-order term $G\,\rho^{\Delta}\,c_j$ captures the bias from the residual unremoved influence. The second-order term $\frac{L}{2}\,\rho^{2\Delta}\,c_j^2$ captures the curvature effect and decays at twice the rate. Because the comparison is to a *same-noise* retraining baseline, the final bound contains no explicit dimension-dependent noise term. If one instead compares to *exact* retraining without noise, an additional smoothing term of order $O(p\sigma_j^2)$ appears.

Minimum Propagation Threshold

The utility bound in Theorem 6.3 decreases monotonically with Δ. The following corollary determines the minimum request-specific propagation interval needed to ensure utility preservation.

Corollary 6.1 (Request-Specific Minimum Propagation Threshold) *Under the conditions of Theorem 6.3, let $y^\star$ be defined by Eq.* (6.9). *Then the utility requirement*

$$\mathbb{E}\big[\hat{R}(\tilde{\boldsymbol{\theta}}^{[\tau]}; \mathcal{D}')\big] - \mathbb{E}\big[\hat{R}(\tilde{\boldsymbol{\theta}}_{-j}^{[\tau]}; \mathcal{D}')\big] \leq \xi$$

is satisfied whenever $\Delta \geq \Delta_{min,j}$, where

$$\Delta_{min,j} = \begin{cases} 0, & c_j \leq y^\star, \\ \left\lceil \dfrac{\ln(c_j/y^\star)}{\ln(1/\rho)} \right\rceil, & c_j > y^\star. \end{cases}$$

Proof Set

$$y = \rho^{\Delta} c_j.$$

Then Theorem 6.3 gives the upper bound

$$Gy + \frac{L}{2}y^2.$$

By construction of $y^\star$, the inequality

$$Gy + \frac{L}{2}y^2 \leq \xi$$

holds if and only if $0 \leq y \leq y^\star$. Therefore it suffices and is necessary that

$$\rho^{\Delta} c_j \leq y^\star.$$

If $c_j \leq y^\star$, then $\Delta = 0$ already suffices. Otherwise, taking logarithms yields

$$\Delta \geq \frac{\ln(c_j/y^\star)}{\ln(1/\rho)},$$

and rounding up gives the claimed threshold. □

The threshold admits a transparent interpretation. The numerator $\ln(c_j/y^\star)$ measures how many orders of magnitude the residual influence must shrink, from its initial scale c_j to the admissible scale $y^\star$. The denominator $\ln(1/\rho)$ is the per-step contraction rate. The threshold is therefore the number of steps needed to accumulate sufficient decay under locally contractive dynamics.

Complexity Analysis

This subsection formalizes the computational and storage costs of the decay-aware unlearning protocol and compares them against three baselines: full retraining, trajectory-based influence correction, and the TVIM-based approach of Chap. 4.

Per-Request Costs

The decay-aware protocol (Algorithm 8) performs two operations at unlearning time: computing the noise standard deviation σ_j from scalar metadata and generating a p-dimensional Gaussian vector. The scalar computation is negligible; Gaussian generation requires $O(p)$ time. The total per-request time complexity is therefore $O(p)$, independent of T, B, and Δ.

Full retraining requires $O(T \cdot B \cdot p)$ operations: repeating the entire training process on $\mathcal{D} \setminus \{z_j\}$. Trajectory-based influence correction requires tracing the influence of z_j through Δ steps, each involving gradient- or Hessian-vector style computation over a mini-batch of size B, for a total of at least $O(\Delta \cdot B \cdot p)$. The TVIM inference algorithm (Chap. 4, Algorithm 3) similarly requires $O(T \cdot B \cdot p)$ for a full-trajectory query.

Storage Overhead

The decay-aware protocol needs only lightweight request metadata: for each deletable sample, either the step index t_{z_j} at which it was used or enough deterministic data-order metadata to recover it, along with global constants such as ρ and G. If stored explicitly for all samples, this is $O(N)$ metadata; no parameter checkpoints, gradient histories, or Hessian information are retained for the unlearning mechanism.

Full retraining requires $O(p)$ storage for the model parameters and $O(N)$ for the dataset but no additional trajectory information. Trajectory-based influence methods typically require storing gradient or Hessian-related information along the training path, totaling at least $O(T(B + p))$. TVIM stores parameter checkpoints and batch indices, requiring $O(T(B + p))$ in its standard implementation and $O(T/C \cdot p + T \cdot B)$ in its checkpoint-based variant (where C is the checkpoint interval).

Latency and Training Overhead

The unlearning latency is $O(p)$ once the maturity condition is met: a single Gaussian vector is generated and added to the parameters. The integration overhead in the training loop consists of tracking pending requests and checking which ones have matured. If requests are stored in a queue or heap keyed by maturity time m_j, the bookkeeping cost is small compared with the $O(B \cdot p)$ cost of each SGD update.

Summary Comparison

Table 6.2 summarizes the complexity comparison across all four methods.

The decay-aware approach achieves the smallest per-request asymptotic cost: $O(p)$ time, independent of the training trajectory length, and lightweight metadata rather than full trajectory storage. The price for this efficiency is the maturity requirement: the guarantee holds only when the propagation interval exceeds the request-specific threshold $\Delta_{\min,j}$. An unlearning request for a recently used data point may therefore need to wait until training has progressed sufficiently for natural decay to reduce its influence to a certifiable level.

Table 6.2 Complexity comparison of unlearning approaches for a model with p parameters, T training steps, batch size B, and propagation interval Δ

Method	Time (per request)	Additional storage	Certified
Retraining	$O(TBp)$	$O(N + p)$	Yes
Trajectory-based influence correction	$O(\Delta Bp)$	$O(T(B + p))$	Approximate
TVIM (Chap. 4)	$O(TBp)$	$O(T(B + p))$	No
Decay-aware (this chapter)	$O(p)$	$O(N)$ metadata or equivalent	Yes

Conclusion

This chapter developed an in-training certified unlearning framework for a simplified but analyzable setting: single-pass SGD-like training with coupled L_2 regularization. Rather than waiting until convergence or restarting from scratch when problematic data is discovered mid-run, the framework exploits local contractive dynamics to let the influence of each data point decay through subsequent optimization steps. After a sufficient number of steps, the residual influence is small enough to be masked by calibrated Gaussian noise, yielding (ε, δ)-certified unlearning relative to a noisy retraining oracle at $O(p)$ cost per request, with no gradient storage, Hessian inversion, or trajectory replay.

The theoretical analysis established three results. First, Theorem 6.1 proved that, under a local contractivity condition, the exact parameter deviation induced by a single data point decays geometrically at rate ρ^{Δ}. Second, Theorem 6.2 showed that isotropic Gaussian noise with standard deviation proportional to the residual influence bound achieves (ε, δ)-certified unlearning relative to a noisy retraining oracle. Third, Theorem 6.3 bounded the excess risk of the unlearned model relative to the same-noise retraining baseline, establishing that utility degradation decays exponentially with the propagation interval. These three results combine to characterize when and how the framework can be applied: the request-specific maturity condition $\Delta \geq \Delta_{\min,j}$ (Corollary 6.1) identifies the minimum number of training steps that must elapse before noise injection is both certifiable and utility-preserving.

The practical message is therefore qualified but useful. When the training dynamics are locally contractive and deletion requests target sufficiently old samples, the required perturbation can be very small and the online overhead is negligible. At the same time, the present analysis does *not* yet cover modern adaptive optimizers such as AdamW, repeated sample exposure, or other large-scale training details that depart from the single-pass SGD-like abstraction. Extending the framework to those settings, together with empirical validation, is left to future work.

References

1. Brown, T., et al.: Language models are few-shot learners. Adv. Neural. Inf. Process. Syst. **33**, 1877–1901 (2020)
2. Touvron, H.: Llama: open and efficient foundation language models. arXiv preprint arXiv:2302.13971 (2023)
3. Gao, L.: The pile: an 800gb dataset of diverse text for language modeling. arXiv preprint arXiv:2101.00027 (2020)
4. Dodge, J.: Documenting large webtext corpora: a case study on the colossal clean crawled corpus. In: Proceedings of the 2021 Conference on Empirical Methods in Natural Language Processing, pp. 1286–1305 (2021)
5. Voigt, P., dem Bussche, A.V.: The EU general data protection regulation (GDPR): a practical guide (2017). https://doi.org/10.1007/978-3-319-57959-7
6. Ginart, A.: Making AI forget you: data deletion in machine learning. In: Advances in Neural Information Processing Systems, vol. 32 (2019)

7. Guo, C.: Certified data removal from machine learning models. In: International Conference on Machine Learning. PMLR, pp. 3832–3842 (2020)
8. Cao, Y., Yang, J.: Towards making systems forget with machine unlearning. In: IEEE Symposium on Security and Privacy, pp. 463–480. IEEE (2015)
9. Bourtoule, L.: Machine unlearning. In: IEEE Symposium on Security and Privacy (SP), pp. 141–159. IEEE (2021)
10. Golatkar, A., Achille, A., Soatto, S.: Eternal sunshine of the spotless net: selective forgetting in deep networks. In: Proceedings of the IEEE/CVF Conference on Computer Vision and Pattern Recognition, pp. 9304–9312 (2020)
11. Neel, S., Roth, A., Sharifi-Malvajerdi, S.: Descent-to-delete: gradient-based methods for machine unlearning. In: Algorithmic Learning Theory. PMLR, pp. 931–962 (2021)
12. Dwork, C.: Calibrating noise to sensitivity in private data analysis. In: Theory of Cryptography Conference, pp. 265–284. Springer (2006)
13. Loshchilov, I., Hutter, F.: Decoupled weight decay regularization. In: International Conference on Learning Representations. https://openreview.net/forum?id=Bkg6RiCqY7 (2019)
14. Krogh, A., Hertz, J.: A simple weight decay can improve generalization. In: Advances in Neural Information Processing Systems, vol. 4 (1991)
15. Dwork, C., Roth, A., et al.: The algorithmic foundations of differential privacy. Found. Trends Theor. Comput. Sci. **9**(3–4), 211–407 (2014)

Chapter 7
Conclusion, Practical Challenges, and Future Directions

Abstract This chapter synthesizes the contributions of the preceding chapters, identifies practical challenges that arise when deploying machine unlearning in wireless network systems, and outlines future research directions. Section "Conclusions" summarizes the book's unified framework connecting data influence measurement to exact and approximate unlearning. Section "Practical Challenges in Wireless Network Deployment" analyzes three categories of wireless deployment challenges: resource-constrained edge computation, federated unlearning under communication constraints, and verification in distributed wireless systems. Section "Future Research Directions" identifies three future research directions: theoretical extensions to the influence decay and certification framework, algorithmic extensions to emerging model paradigms, and integration of machine unlearning into next-generation wireless architectures, including O-RAN intelligent controllers and AI-native sixth-generation networks.

Conclusions

Machine learning systems encode the influence of their training data into model parameters through optimization trajectories. This book has developed a unified framework for measuring, understanding, and removing that influence, progressing from theoretical foundations through practical algorithms that operate at the scale of modern large models.

Data Influence as the Theoretical Foundation

The ability to remove a data point's influence from a trained model requires, as a prerequisite, the ability to measure that influence precisely. Chapters 2, 3, and 4 developed a progressively more refined understanding of how individual data points shape the optimization trajectory and the resulting model parameters.

J. Xu and X. Jia, *Machine Unlearning*, Wireless Networks,
https://doi.org/10.1007/978-3-032-30978-5_7

Chapter 2 surveyed the classical landscape of data influence analysis, from leave-one-out retraining and influence functions through game-theoretic valuation (Shapley, Banzhaf) and training trajectory methods (TracIn, GraNd). The survey revealed five structural limitations shared by these classical approaches: computational scalability barriers, violated assumptions in non-convex settings, the inability to capture dynamic (as opposed to static) influence, training order bias in gradient-based methods, and the empirical discrepancy between influence function estimates and true leave-one-out effects. These limitations motivated the development of the two influence measurement frameworks that constitute the theoretical core of the book.

Chapter 3 addressed the temporal blindness of classical methods by introducing LiveVal, a real-time data valuation framework that assesses each data point's contribution to the optimization trajectory as training proceeds. LiveVal computes a normalized distance-reduction score at each training step, measuring whether a data point's gradient moves the parameters closer to or further from an adaptive reference point. The adaptive reference mechanism dynamically adjusts the lookahead window based on the training loss change rate, expanding during volatile phases and contracting near convergence. Three theoretical properties (Theorems 3.1 through 3.3) established that LiveVal valuations are directionally aligned with true influence, bounded within $[-1, 1]$, stable under stochastic training noise, and temporally fair in the sense that data points with identical gradient properties receive identical valuations regardless of when they appear in the training sequence. The experimental validation demonstrated a 180-fold speedup over leave-one-out evaluation with competitive detection accuracy, scaling to models with 24 million parameters.

Chapter 4 extended the analysis from instantaneous to cumulative influence through the TVIM (Time-Varying Influence Measurement) framework. TVIM introduced the propagator matrix $Z_t = I - \eta_t H^{[t]}$ whose products govern how a data point's parameter deviation evolves across subsequent optimization steps. The two-phase design (trajectory recording during training followed by retrospective backward propagation of query vectors) enables the computation of data influence over arbitrary sub-intervals $[t_1, t_2]$ of the training process, revealing *when* during training each data point exerted its effect. The error bound of Theorem 4.1 established that the estimation accuracy degrades at most exponentially with the window length under non-convex conditions, controlled by the learning rate and Hessian smoothness. Four query instantiations (loss value, prediction change, feature importance, and parameter importance) provided a toolkit for different analytical needs. The experimental results demonstrated that TVIM consistently outperforms influence functions across all settings, with the advantage most pronounced in non-convex deep networks, and that late-epoch influence windows are particularly effective for detecting mislabeled data.

Together, LiveVal and TVIM provide complementary capabilities: LiveVal enables proactive, real-time identification of harmful data during training (before influence becomes deeply embedded), while TVIM enables retrospective analysis that localizes influence to specific training windows (guiding targeted removal). This combination of proactive detection and retrospective localization underpins the unlearning algorithms developed in Part II.

Unlearning Algorithms as the Practical Framework

Part II translated the theoretical insights of Part I into exact and approximate unlearning approaches.

Chapter 5 developed LMEraser, an exact unlearning method for large-scale vision models that exploits the two-phase structure of modern model training, pre-training on public data followed by tuning on private data. LMEraser freezes the backbone after pre-training, ensuring that private data influences only the prompt and classifier head parameters. Private data is then partitioned into diversity-aware clusters using hierarchical agglomerative clustering on backbone features, with a dedicated visual prompt and classifier head trained for each cluster. Unlearning a data point requires re-optimizing only the affected cluster's prompt and classifier head, achieving exact removal while reducing the volume of affected training data by approximately 100-fold and the number of affected parameters by approximately 1000-fold relative to full retraining. The theoretical analysis established that the expected efficiency gain is independent of the request arrival rate, the batching period, and the cluster size distribution, depending only on the ratio of single-cluster to full-model retraining costs. The experimental validation across five private datasets and four backbone architectures confirmed that LMEraser preserves model utility while reducing per-request unlearning time from hours to seconds.

Chapter 6 developed a certified unlearning framework that operates during training by exploiting a structural property of regularized optimization. Building on the propagator matrix analysis of Chap. 4, the chapter established that weight decay causes the spectral norm of every propagator matrix to be uniformly bounded below one ($\|Z_t\|_2 \leq \rho < 1$), so that the influence of any single data point decays geometrically through subsequent optimization steps at rate $\rho_{\text{eff}}^{\Delta}$, where Δ is the number of elapsed steps (Theorem 6.1). This exponential decay transforms the unlearning problem: rather than explicitly computing and subtracting influence (which requires stored gradients, Hessian information, and trajectory replay), the framework injects isotropic Gaussian noise calibrated to the residual influence bound, achieving (ε, δ)-certified unlearning (Theorem 6.2) at $O(p)$ per-request cost. The utility analysis (Theorem 6.3) showed that the excess risk relative to the retrained baseline decays exponentially with Δ and is independent of the model dimension, since both the unlearned and retrained models receive identically distributed noise. The maturity condition $\Delta \geq \Delta_{\text{min}}$ (Corollary 6.1) identified the minimum number of training steps that must elapse before noise injection is both certifiable and utility-preserving, with representative large-model parameters yielding $\Delta_{\text{min}} \approx 3.5 \times 10^5$ steps, a modest fraction of a typical multi-million-step training run. The resulting computational advantage over full retraining exceeds 10^9 for typical configurations.

Practical Challenges in Wireless Network Deployment

Translating the theoretical and algorithmic results of this book into deployed unlearning systems for wireless networks requires addressing practical challenges that arise from the distinctive resource constraints, distributed architectures, and non-stationary operating environments of modern wireless infrastructure. This section analyzes three categories of such challenges: resource-constrained edge computation (section "Resource-Constrained and Heterogeneous Edge Computation"), federated unlearning under communication constraints (section "Federated Unlearning Under Communication Constraints"), and verification in distributed wireless systems (section "Verification in Distributed Wireless Systems").

Resource-Constrained and Heterogeneous Edge Computation

The computational profiles of the unlearning methods developed in this book are favorable relative to full retraining, but wireless deployments impose absolute resource ceilings that translate relative efficiency into hard feasibility questions. Edge servers, small-cell base stations, RIC-hosted xApps, and user equipment operate under stringent Size, Weight, and Power (SWaP) budgets that are orders of magnitude tighter than the datacenter environments in which most unlearning algorithms are validated [1, 2]. A near-real-time RIC xApp executing a resource management policy must complete all computation, including any unlearning operation, within the control loop's time budget, typically on the order of seconds to minutes [3]. This constraint transforms the unlearning problem from a question of computational efficiency into a question of absolute feasibility.

The decay-aware noise injection framework (Chap. 6) is architecturally well-suited to edge deployment due to its $O(p)$ per-request cost, requiring only the generation and addition of a Gaussian noise vector with no gradient storage, Hessian computation, or trajectory replay. However, for models deployed at the device level (on-device personalization models on user equipment or IoT sensors), the model may be too small and the training run too short for the maturity condition $\Delta \geq \Delta_{\min}$ to be satisfied within a single training cycle. In this regime, the residual influence has not decayed sufficiently for noise injection to be both certifiable and utility-preserving, and the framework falls back to the more expensive alternatives. LMEraser (Chap. 5) avoids the maturity requirement by providing exact removal through cluster re-optimization, but the memory overhead of maintaining the clustering structure, the per-cluster prompt parameters, and the backbone feature representations may exceed the capacity of resource-constrained edge devices.

A wireless deployment typically spans multiple tiers of the network hierarchy: device-level models for local personalization, edge-level models for cell-site optimization hosted on RIC infrastructure, and cloud-level models aggregated from federated training across the network. Each tier has different computational budgets, memory constraints, and latency requirements, and the appropriate unlearning method differs across tiers. Designing a tiered unlearning architecture that selects the appropriate method for each network tier, while maintaining consistent privacy guarantees across the hierarchy, is an open systems design problem. The challenge is compounded by the need for cross-tier coordination: when a user exercises a deletion right, the request must propagate to every tier that hosts a model trained on that user's data, and the unlearning must be executed consistently across all tiers despite their heterogeneous computational capabilities.

A related concern arises from the memory footprint of influence measurement at the edge. Developing memory-efficient variants of TVIM that exploit the structure of edge model training (smaller models, shorter training runs, potentially quantized or pruned parameter representations) is an important direction for making influence-guided unlearning practical at the wireless edge.

Federated Unlearning Under Communication Constraints

Federated learning is the predominant training paradigm for wireless edge intelligence [4, 5], and it introduces a fundamental challenge for unlearning that has no counterpart in centralized settings: the communication cost of coordinating removal across distributed clients.

In federated wireless learning, training data are partitioned across K clients (devices, base stations, IoT sensors) that do not share raw data. Each client's contribution is interleaved with others across many rounds, so a single client's influence is dispersed throughout the entire sequence of global model updates rather than confined to a contiguous training window. When a client withdraws consent, the ideal operational goal is to remove that client's influence without requiring all remaining clients to re-participate in extra rounds beyond the standard protocol. Achieving this goal is technically difficult: exact unlearning typically requires either partitioning the federation into disjoint sub-models (reducing effective participation per model and potentially degrading utility) or re-executing the federated protocol without the departing client's updates (incurring essentially the full communication cost of training).

The decay-aware noise injection framework (Chap. 6) suggests a promising direction, but its calibration in FL requires new theory. The effective contraction rate ρ_{eff} depends on the aggregation rule, local steps per round, client sampling, and compression (quantization/sparsification/top-k). Under heterogeneity, local updates can amplify client-specific deviations (client drift [6]), potentially violating uniform contraction assumptions. Closing this gap requires federated-specific influence accounting that (a) bounds per-client contributions under realistic participation and compression, (b) characterizes when aggregation induces contraction in spite of drift, and (c) produces server-side certificates without storing per-client gradient histories.

Verification in Distributed Wireless Systems

Wireless operators face compliance obligations under general data protection regimes (GDPR, CCPA) and sector-specific telecommunications rules (e.g., the ePrivacy Directive [7]), and these obligations increasingly require *demonstrable* handling of deletion requests. In distributed wireless systems, verification is harder than in centralized deployments for two reasons: models are replicated across tiers (device/edge/cloud), and training is often federated or otherwise distributed, so removal must be both *complete across locations* and *consistent across replicas*.

For certified approximate unlearning, the (ε, δ) guarantee (Definition 1.17) provides a formal statement of indistinguishability relative to retraining under a fixed protocol. In practice, an auditor may also demand a *chain of evidence* that links (i) the deletion request, (ii) the set of models affected (provenance), (iii) the execution of the unlearning procedure at each tier, and (iv) post-hoc validation results. This creates an engineering requirement that is separate from the learning theory: operators need deletion routing, provenance metadata, and audit logs that are reliable across distributed components.

A second challenge is *assumption accountability*. Certified guarantees depend on modeling assumptions (e.g., bounded gradients and stability of training dynamics). Current frameworks typically state these assumptions but do not provide deployment-grade mechanisms to monitor or justify them for a specific system configuration. Developing "certificate-supporting telemetry", including lightweight checks, conservative bounds, and configuration constraints that make the assumptions auditable, is essential for real deployments.

Finally, empirical verification remains necessary but incomplete. Membership inference attacks, distributional distance measures, and task-specific forget tests (surveyed in Chap. 2) provide practical signals, yet no single metric captures all leakage channels: passing one attack suite does not rule out residual semantic memorization detectable by stronger probes [8]. A robust wireless verification stack should therefore combine (a) formal certificates when available, (b) multi-metric empirical auditing, and (c) system-level provenance and logging that demonstrate end-to-end completeness across all deployment tiers.

Future Research Directions

The preceding section identified gaps between the current state of the art and the requirements of wireless network deployment. This section identifies three research directions that extend the scope of machine unlearning beyond the settings studied in this book.

Theoretical Extensions

The theoretical results of this book establish influence measurement and certified unlearning under three assumptions that each invite relaxation.

The first is the *single-pass training assumption*. The decay analysis of Chap. 6 assumes each data point appears in exactly one mini-batch. Under multi-epoch training, a data point z_j contributes gradients at multiple steps $t_{z_j}^{(1)}, \ldots, t_{z_j}^{(E)}$, and its residual influence becomes a superposition:

$$\|\Delta\boldsymbol{\theta}_{-j}^{[\tau]}\| \leq \sum_{e=1}^{E} \rho_{\text{eff}}^{\tau - t_{z_j}^{(e)} - 1} \cdot c_e. \tag{7.1}$$

Deriving tight bounds and maturity conditions under general multi-pass schedules is a natural extension with immediate practical relevance, since most production pipelines employ multiple epochs.

The second is the *independent composition assumption*. The basic composition theorem yields a privacy budget growing linearly with the number of sequential requests. Advanced composition [9] and Rényi divergence accounting [10] improve this, but assume independent mechanisms. When sequential noise injections interact through subsequent training dynamics, tighter composition theorems that exploit the structure of interleaved noise injection and SGD could substantially extend the model's operational lifetime before a budget refresh is required.

The third is the *absence of lower bounds*. The current results provide upper bounds on noise magnitude, maturity threshold, and utility degradation, but no corresponding lower bounds. Information-theoretic lower bounds would clarify whether the exponential decay rate achieved by weight decay is optimal or whether alternative regularization strategies could achieve faster influence attenuation.

Algorithmic Extensions to Emerging Model Paradigms

The methods developed in this book target supervised learning with clearly defined data points and gradient-based parameter trajectories. Three emerging paradigms challenge these assumptions.

Foundation models and generative AI. Self-supervised pre-training defines "data points" by tokenization and context windowing rather than natural sample boundaries, making the forget set ambiguous. RLHF fine-tuning mediates influence through reward models and policy gradient estimators. For generative models (diffusion, VAE, GAN), unlearning must ensure the output distribution no longer "contains" the forgotten data while preserving generalization to similar content, a conceptual challenge absent in discriminative settings.

Structured learning and non-i.i.d. data objects. In graph learning, the natural "forget targets" are often nodes, edges, or subgraphs, and influence propagates not only through time (optimization steps) but also through graph connectivity and neighborhood sampling. Although GNNs are typically trained with gradient-based optimizers, the dependence structure induced by message passing and minibatch neighborhood expansion changes both influence tracing and what it means to "remove" a data object. In reinforcement learning, influence is mediated through temporal-difference updates, replay buffers, and policy improvement steps, and the forget set may correspond to trajectories or experience segments rather than i.i.d. samples. Both paradigms are increasingly deployed in wireless systems for resource management and spectrum access [11]. Extending influence measurement and certified removal to these structured settings would significantly broaden the applicability of unlearning in wireless networks.

Adaptive noise injection. The current framework injects isotropic noise uniformly across all parameters, but data influence is typically localized to specific layers or parameter groups [12, 13]. Combining influence localization with targeted noise injection could reduce the total noise required for certification, improving utility preservation while maintaining the (ε, δ) guarantee.

Integration Into Next-Generation Wireless Architectures

Deploying machine unlearning in next-generation wireless systems requires advances in protocol-level integration, network architecture design, and supporting ecosystem infrastructure.

Communication-efficient federated unlearning. Existing federated unlearning methods require multiple communication rounds or per-client gradient histories at the server [14, 15]. Extending the noise injection framework to federated averaging, by bounding each client's contribution and exploiting aggregation-induced contraction to calibrate server-side noise, could enable certified removal with no additional communication beyond the standard training protocol.

O-RAN integration. The O-RAN architecture's standardized interfaces (A1, E2, O1) provide natural integration points for unlearning workflows, but doing so requires explicit support for provenance and deletion routing. Open questions include extending management-plane workflows (e.g., via O1) for deletion request propagation, maintaining per-model provenance metadata in the RIC, and ensuring that unlearning actions respect near-real-time control-loop budgets (typically sub-second, e.g., 10 ms to 1 s for near-RT functions). The $O(p)$-cost noise injection framework is particularly attractive for tiers where parameters can be updated quickly, whereas exact modular retraining may be better aligned with non-real-time orchestration where longer latencies are acceptable.

AI-native 6G networks. When trained models replace hand-crafted algorithms across the protocol stack [16], unlearning becomes a network management primitive. This setting introduces cross-layer influence propagation (jointly trained models sharing representations across protocol layers), network-wide consistency requirements (unlearning must execute consistently across all base stations sharing a model), and spectrum-aware constraints (parameter perturbations must not violate regulatory spectrum masks during the transition).

Privacy-preserving influence measurement. LiveVal and TVIM require per-sample gradients, which in the federated setting may reveal client data. Privacy-preserving variants operating on differentially private or securely aggregated gradients, potentially leveraging TEEs on modern edge hardware, would enable influence-guided unlearning without compromising client privacy.

Ecosystem. Wireless-specific unlearning benchmarks (incorporating realistic channel models, federated protocols, non-stationary data streams), regulatory clarity on whether “erasure” requires removing model-parameter-level influence, and integration into wireless simulation platforms (ns-3, O-RAN Software Community) would collectively lower the barrier to adoption.

Closing Remark

Machine unlearning transforms the relationship between data and models from one-way inscription into a controllable lifecycle: data can be incorporated through training, audited through influence measurement, and removed through unlearning with guarantees calibrated to operational constraints. This book developed (i) measurement tools that make influence observable in modern, stochastic, non-convex training, and (ii) removal mechanisms spanning exact modular retraining and certified approximate unlearning via decay-aware noise injection. It also mapped these tools to the distinctive requirements of wireless deployments, including multi-tier distribution, tight edge budgets, non-stationarity, and regulatory auditing.

Realizing deployed unlearning systems that are scalable, verifiable, and compatible with real-time wireless control will require coordinated progress across learning theory, systems engineering, privacy-preserving computation, and wireless architecture standardization. The framework developed in this book provides a concrete starting point for that interdisciplinary effort.

References

1. Chen, J., Ran, X.: Deep learning with edge computing: a review. Proc. IEEE **107**(8), 1655–1674 (2019)
2. Li, E., et al.: Edge AI: on-demand accelerating deep neural network inference via edge computing. IEEE Trans. Wirel. Commun. **19**(1), 447–457 (2019)
3. Polese, M., et al.: Understanding O-RAN: architecture, interfaces, algorithms, security, and research challenges. IEEE Commun. Surv. Tutor. **25**(2), 1376–1411 (2023)

4. McMahan, B., et al.: Communication-efficient learning of deep networks from decentralized data. In: Artificial Intelligence and Statistics. PMLR, pp. 1273–1282 (2017)
5. Lim, W.Y.B., et al.: Federated learning in mobile edge networks: a comprehensive survey. IEEE Commun. Surv. Tutor. **22**(3), 2031–2063 (2020)
6. Karimireddy, S.P., et al.: Scaffold: stochastic controlled averaging for federated learning. In: International Conference on Machine Learning. PMLR, pp. 5132–5143 (2020)
7. Bond, R.: The EU e-Privacy directive and consent to cookies. Bus. Law. 215–223 (2012)
8. Carlini, N., et al.: Quantifying memorization across neural language models. In: The Eleventh International Conference on Learning Representations (2022)
9. Dwork, C., Roth, A., et al.: The algorithmic foundations of differential privacy. Found. Trends Theor. Comput. Sci. **9**(3–4), 211–407 (2014)
10. Mironov, I.: Rényi differential privacy. In: IEEE 30th Computer Security Foundations Symposium (CSF), pp. 263–275. IEEE (2017)
11. Mao, Q., Hu, F., Hao, Q.: Deep learning for intelligent wireless networks: a comprehensive survey. IEEE Commun. Surv. Tutor. **20**(4), 2595–2621 (2018)
12. Meng, K., et al.: Locating and editing factual associations in GPT. In: Oh, A.H., et al. (eds.) Advances in Neural Information Processing Systems. https://openreview.net/forum?id=-h6WAS6eE4 (2022)
13. Meng, K., et al.: Mass-editing memory in a transformer. In: The Eleventh International Conference on Learning Representations. https://openreview.net/forum?id=MkbcAHIYgyS (2023)
14. Wu, C., et al.: A federated graph neural network framework for privacy-preserving personalization. Nat. Commun. **13**(1), 3091 (2022)
15. Liu, G., et al.: Federaser: enabling efficient client-level data removal from federated learning models. In: IEEE/ACM 29th International Symposium on Quality of Service (IWQOS), pp. 1–10. IEEE (2021)
16. Letaief, K.B., et al.: The roadmap to 6G: AI empowered wireless networks. IEEE Commun. Mag. **57**(8), 84–90 (2019)

FSC
www.fsc.org
MIX
Papier | Fördert
gute Waldnutzung
FSC® C083411

Kolibri 360 GmbH
Ferdinand-Jühlke-Straße 7
99095 Erfurt, Deutschland
produktsicherheit@kolibri360.de